DESIGNING
SIMPLE
INSTRUCTIONS
FOR MEDICAL PRODUCTS

Designing Simple Instructions for Medical Products motivates and guides readers to create excellent instructions for medical products. The book is just what the doctor ordered to craft instructions that enhance usability and safety for products used in hospitals, clinics, laboratories, and at home. For doctors, nurses, therapists, technicians, and laypeople, clear and simple instructions can be lifesavers, figuratively and literally.

When medical product developers disregard instructions and underinvest in their development, the result is user confusion, dissatisfaction, and an increased potential for use errors. In contrast, well-designed instructions, implementing some or all tips in this book, can (1) help ensure a product's safe and effective use and (2) drive greater commercial success by satisfying users, strengthening brand identity, and simplifying product use.

This book provides actionable principles and advice in a readable, artful, and entertaining manner. Inside, you'll find practical guidance on every stage of the instructional development process, including page layout, information hierarchy, writing style, illustration style, and production values.

SPECIAL BOOK FEATURES INCLUDE:

- An approachable, down-to-earth explanation of how to translate user needs and preferences into instruction design requirements.
- Visual explanations of key design principles to aid understanding and implementation.
- Detailed guidance on creating instructions that reduce the use-related risks of operating a specific medical product.
- Sample, exemplary instructions for medical devices, pharmaceutical combination products, and in vitro diagnostic devices.
- Advice on critical yet often overlooked details, such as paper thickness, instruction sheet folding, font size selection, and much more.

The book is essential reading for technical writers, communications designers, human factors specialists, industrial designers, graphic designers, and clinical specialists involved in medical product development.

DESIGNING
SIMPLE
INSTRUCTIONS
FOR MEDICAL PRODUCTS

Michael Wiklund

Jonathan Kendler

Katelynn Larson

CRC Press
Taylor & Francis Group
Boca Raton London New York

CRC Press is an imprint of the
Taylor & Francis Group, an **informa** business

Designed cover image: Jonathan Kendler

First edition published 2025
by CRC Press
2385 NW Executive Center Drive, Suite 320, Boca Raton FL 33431

and by CRC Press
4 Park Square, Milton Park, Abingdon, Oxon, OX14 4RN

CRC Press is an imprint of Taylor & Francis Group, LLC

ISBN: 9781032722900 (hbk)

ISBN: 9781032722894 (pbk)

ISBN: 9781032722962 (ebk)

ISBN: 9781040391143 (eBook+)

DOI: 10.1201/9781032722962

Typeset in Barlow
by Jeremy Tribby

Contents

Acknowledgements

Professional and historical acknowledgments

To start, we must acknowledge the pre-existing, large body of knowledge about designing and writing instructions. Instructions have existed for centuries in various forms, and you might wonder what more can be said about producing good ones in view of the existing literature and related resources.

Instructional documents have existed for thousands of years. Early examples include the above Babylonian tablet (left), which is dated from 1386 - 1186 BCE and includes cuneiform instructions for making red glass. Other early artifacts, like the the Ancient Egyptian Book of the Dead (right) from around 1486 BCE, demonstrate the use of instructional illustrations and visual design techniques to better communicate their contents.

Well, we hope this book offers something new, particularly regarding designing instructions for medical products. These products present special use models as compared to such challenges as assembling furniture or driving a car, for example. However, we have built on a base of knowledge, explaining why our book presents a lot of work by others along with our own, and includes many references plus a list of resources (see "Resources" on page 211.

Personal acknowledgements

Michael Wiklund

I thank my wife Amy for continuing to support my special projects including writing this book. I broadly thank my work colleagues; folks who have generously shared their know-how on various topics including designing and writing instructions. This includes my most recent colleagues at UL who continually impress me with their great designs for medical product instructions.

Jonathan Kendler

I thank my family for the support and understanding they provided over the many months during which I worked on this book. My heartfelt thanks go to my wife, Hanna, who offered motivation and encouragement throughout the book writing process, and to my three wonderful kids, who were a constant source of inspiration, energy, and occasional distractions. Professionally, I have tremendous gratitude for my co-author, former business partner, and friend, Michael Wiklund, whose insights and guidance have always been invaluable. Finally, I extend my appreciation to my various clients over the years, whose interesting projects, trust, and collaboration have enabled me to develop many of the best practices we share in this book.

Katelynn Larson

I first would like to thank my wife Tracy for providing the space, time, and support for me to contribute to this book. I also want to thank my co-authors for inviting me the join them on this expedition, as well as my colleagues from whom I have learned much about human factors and designing for medical devices. Finally, I would like to thank Winnie the dog for the company and persistent reminder to go for a walk once in a while.

About the authors

Michael Wiklund

Michael Wiklund served for many years as general manager of the human factors engineering (HFE) practice at UL Solutions (also called Underwriters Laboratories) which does business as Emergo by UL. The large consulting practice has teams based in the USA, the UK, The Netherlands, and Japan. His current role is to help focus and expand UL's work in the life sciences. He is a Corporate Fellow at UL. Previously, he was co-founder and president of Wiklund Research & Design, a human factors engineering and design consultancy that UL acquired in 2012.

A Certified Human Factors Professional, he has over four decades of experience in human factors engineering, much of which has focused on medical technology development; optimizing hardware and software user interfaces as well as user documentation. He is author, co-author, or editor of several books on human factors, including *Usability Testing of Medical Devices*, *Handbook of Human Factors in Medical Device Design*, *Medical Device Use Error – Root Cause Analysis*, and *User Interface Requirements for Medical Devices*. He is one of the primary authors of today's most pertinent standards and guidelines on human factors engineering of medical devices: *AAMI HE75* and *IEC 62366*.

In additional to serving as a Principal Consultant in UL Solutions' human factors engineering practice, he is a Professor of the Practice at Tufts University where he has taught graduate courses on HFE since 1987, including applying HFE in medical technology development. He founded the university's online certificate program focused on applying human factors in medical devices and systems development.

In 2018, he received the Human Factors and Ergonomics Society's A. R. Lauer Safety Award and serves as the society's Vice Chair – Industry Advisory Panel.

Jonathan Kendler

Jonathan Kendler is a user interface designer and human factors engineer specializing in medical technology. Jonathan has over 20 years of experience working on a wide range of medical devices including dialysis machines, patient monitors, infusion pumps, and glucose monitors, as well as on the accompanying instructions of such products.

Together with Michael Wiklund, Jonathan co-founded Wiklund Research & Design, a human factors engineering and design consultancy in 2004, which was acquired by UL (Underwriters Laboratories) in 2012. From 2012 to 2016, Jonathan lead UL's European human factors engineering team based in The Netherlands. From 2016 to 2018 Jonathan lead UL's Life & Health Sciences digital solutions group. Since 2018, Jonathan has worked as principal consultant for Curiolis, a small design consultancy that focuses on medical technology.

In addition to his consulting work, Jonathan has authored numerous articles on user interface design and human factors engineering, as well as co-authored two books (Usability Testing of Medical Devices [2015] and Designing for Safe Use [2019]). Jonathan also teaches university courses about user interface design at Tufts University and has delivered workshops and lectures about user interface design and human factors engineering throughout Asia, Europe, and North America.

Katelynn Larson

Katelynn Larson is a technical writer with experience developing instructional and informational documents as well as other collaterals for the insurance, software, and medical device industries. In her current role at UL Solutions, she supports the Human Factors Research & Design team with conveying technical information regarding regulatory requirements, industry trends, and research methodologies to general and industry audiences. Previously, she developed reference documents and user manuals in collaboration with multidisciplinary teams describing the assumptions and processes associated with actuarial and catastrophe risk models to enable understanding and use by insurance underwriters.

Limitations of our advice

Crafting effective instructions is both a science and an art. It relies on a science-based understanding of human perception and cognitive abilities and requires a rigorous evaluation of people's abilities to interpret and follow instructions. Simultaneously, it demands creative decisions and aesthetic considerations to inspire and motivate users to operate products correctly.

We have infused our writing with empirical and scientific evidence wherever possible. However, this book ultimately contains advice based on our personal experience designing and evaluating hundreds of instructional documents for medical products. Through this experience, we have developed an understanding of the characteristics and processes that tend to make instructions effective and user-friendly. We hope you find this advice helpful.

Our advice focuses primarily on the design of effective instructions and is not intended to be an exhaustive primer on international regulations for medical product instructions. Different countries have different regulations regarding the technical, legal, and safety details that must be included in medical product instructions. Specifying these requirements in detail is beyond the scope of this book. That said, we expect the advice within this book to be generally consistent with contemporary regulatory requirements and expectations.

As the name of this book implies, our advice focuses on creating relatively simple instructions. We believe that instructions for all medical products can and should be as simple as possible, avoiding complexity for its own sake. We recognize that some products are inherently more complex than others and will require more detailed instructions. For example, a robotic surgical system's user manual will include significantly more content than a quick reference guide for a pen-injector. However, we believe both documents would benefit equally from applying the design practices and guidelines described in this book.

Summary disclaimers

This book was prepared by the authors in their personal capacities. The opinions expressed herein are the authors' own and do not reflect the views of their employers.

Any similarity to actual persons, living or dead, is purely coincidental. Hypothetical examples of medical products and instructions reflect a broad base of professional experience and are not attributable to any single product. Product details have been described in generic terms.

CHAPTER ONE
INTRODUCTION

1

Introduction

Welcome to this book about designing simple instructions. We wrote it for people like you who want to design excellent instructions for medical products.

Perhaps we should instead say we wrote this book for people who <u>must</u> design excellent instructions for medical products. In our view, it's a must because medical product users frequently perform tasks with no room for mistakes, which are also termed "use errors." After all, mistakes with some medical products could prove injurious or deadly. And these mistakes could be triggered by confusing instructions. Some awkward wording, a misleading graphic, or the haphazard arrangement of information can all cause trouble.

Writing instructions for medical products brings with it a solemn responsibility. Your work can save lives or put them in jeopardy. This is not hyperbole. When somebody commits a use error while operating a medical product (e.g., glucose meter, automated external defibrillator, intravenous infusion pump, or heart lung machine), the outcome can be fatal. People have died due to medication underdoses and overdoses, air being pumped into blood lines and causing an embolism, and infection from contaminated devices that were not sterilized properly. Excellent instructions are one way to reduce the risk of injury and death due to use errors.

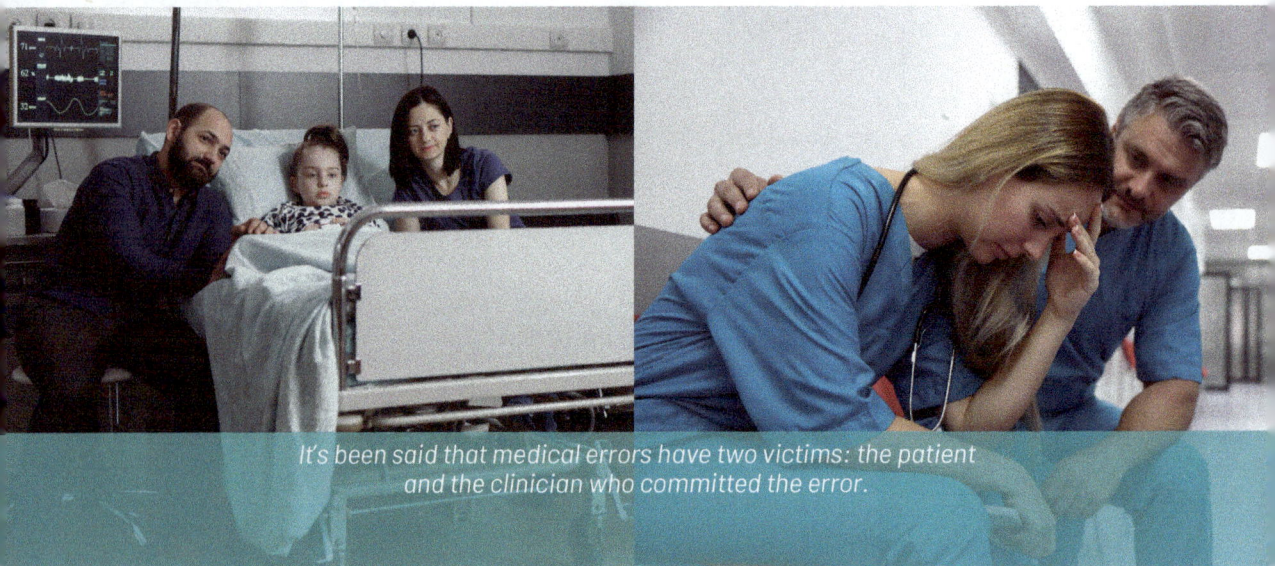

It's been said that medical errors have two victims: the patient and the clinician who committed the error.

Many errors are built into existing routines and devices, setting up the unwitting physician and patient for disaster. And, although patients are the first and obvious victims of medical mistakes, doctors are wounded by the same errors: they are the second victims.[*]

---

[*] Albert W. Wu. Medical error: the second victim. BMJ. 2000 Mar 18; 320(7237): 726–727. https://www.ncbi.nlm.nih.gov/pmc/articles/PMC1117748/

Of course, instructions are not the only way to mitigate use-related risk. In fact, they rest on a lower tier of effectiveness along with training. The most effective way to reduce use-related risk is to produce inherently safe medical products; products that are invulnerable to use errors. However, that's a difficult goal to achieve given the practicalities of many types of medical devices. Yes, you can design a pen-injector to cover the needle after an injection to avoid a needlestick injury. But, vulnerabilities remain, such as not cleaning the injection site or injecting the wrong part of the body. Excellent instructions can facilitate safe and effective interactions with medical products. It's a fact. We've seen proof over years of observing people use medical products in both real-world and simulated settings (e.g., in usability tests).

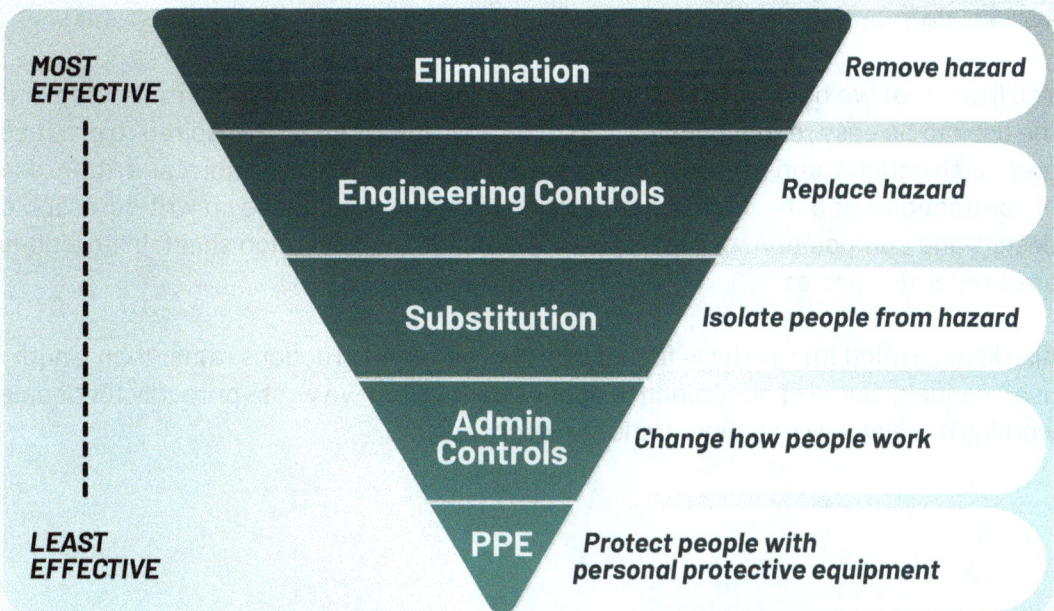

MOST EFFECTIVE	Elimination	Remove hazard
	Engineering Controls	Replace hazard
	Substitution	Isolate people from hazard
	Admin Controls	Change how people work
LEAST EFFECTIVE	PPE	Protect people with personal protective equipment

Instructions are a form of administrative control that can change the way people work. They fall low in the hierarchy but remain an important risk control measure.

Therefore, cast aside suggestions from a disbelieving minority (we hope it's a minority by now) that instructions are useless and that people ignore instructions. As we recognize in "Users might have low expectations for instructions" on page 19, some people will indeed disregard instructions. But many people will depend on them heavily at times, including when learning to use a medical product and when they run into trouble.

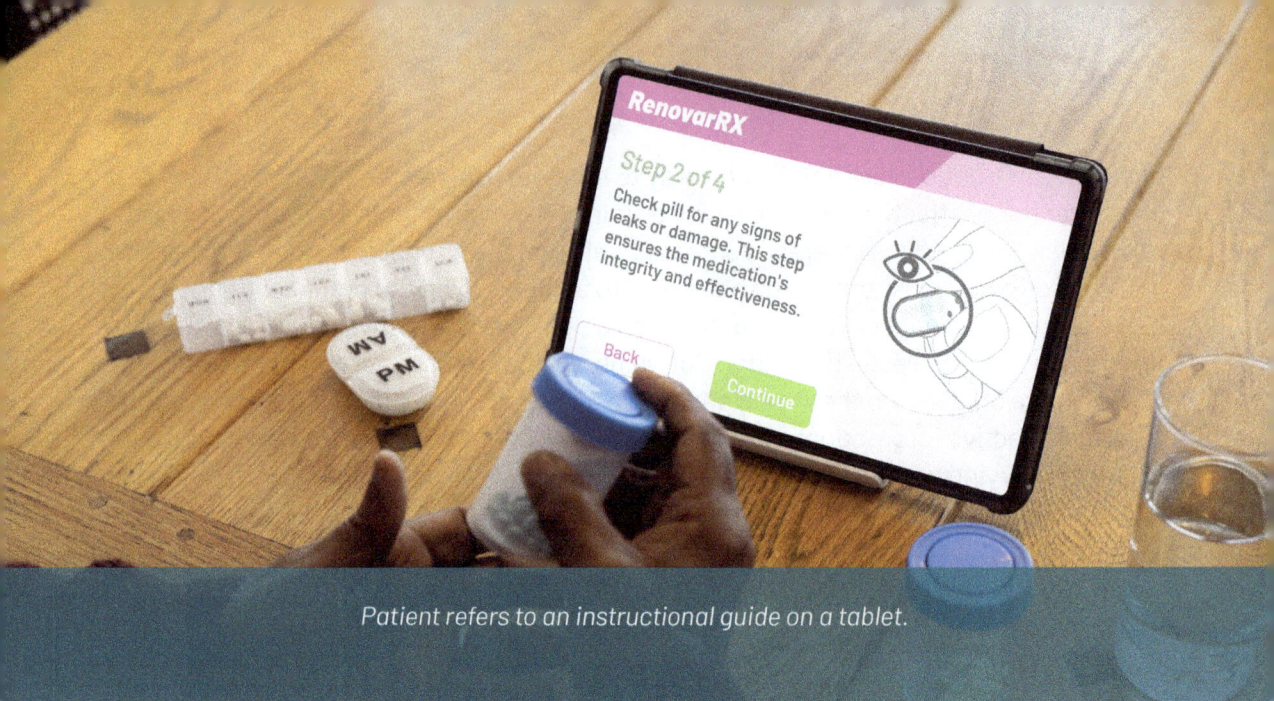

Patient refers to an instructional guide on a tablet.

You'll see that we have written this book in a mostly conversational tone. We want the book to be easy reading and for you to be well-informed by it, inspired to produce great instructions, and maybe even amused. Please forgive our informal writing, use of contractions, and the occasional violation of our own guidance on writing, graphic design, and composition. But this is a book and not an instruction sheet. Instructions are a separate species.

Also keep in mind that we have focused on short-form instructions rather than lengthy user manuals. Some of our guidance applies to both, but we wrote primarily for people working on relatively short instructions, which include:

Quick reference guides

Single-sheet instructions

Package inserts

Online help screens

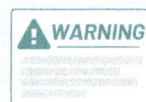

Warnings and notices

Another thing about this book: it has numbered sections. Though "section" is a bit of a stretch. Perhaps "vignette" would have been a more appropriate term, but that sounds a bit too pretentious for our tastes. We've tried to be clear and concise, striving to limit each section to just a few, well-illustrated pages. In the process of writing the book, we kept reducing sections down to their minimum. We took a lesson from the adage, "If I had more time, I would have written a shorter letter," and spent the time to write less.

Accordingly, we took some of the "medicine" that we dispense in this book. Then again, in "Facilitate translation" on page 87 we warn against using idioms and other linguistic tricks that will not translate well (like the one we used in the prior sentence). Well, well.

On that note, let's get into the technical content.

SIDEBAR
"If I had more time, I would have written a shorter letter."

The first known instance of this sentiment in the English language was a sentence translated from a text written by the French mathematician and philosopher Blaise Pascal.

The French statement appeared in a letter in a collection called "Lettres Provinciales" in the year 1657:

Je n'ai fait celle-ci plus longue que parce que je n'ai pas eu le loisir de la faire plus courte.

Here is one possible modern day translation of Pascal's statement:

I have made this longer than usual because I have not had time to make it shorter.

Note that the term "this" refers to the letter itself.

2
What constitutes instructions?

Among people working in the medical industry, few have the time or energy to say the entire phrase "instructions for use." They usually refer to the acronym "IFU," and they may be referring to one of several kinds of resources.

Documents that a person or organization calls an IFU can vary from a one-page leaflet to a multi-page booklet or book. Such documents may take a physical form, electronic form (i.e., an eIFU), or both.

To complicate matters, the term IFU is often used synonymously with many other terms, including the following:

- Directions for use
- Information for safety
- Instructional leaflet
- Instructions
- Job aid
- Owner's manual
- Package insert
- Patient insert
- Procedure guide
- Quick reference guide
- Quick start guide
- User guide
- User manual

For the purposes of this book, we consider it unnecessary to establish strict boundaries for documents qualifying as an IFU versus not. However, here is a list of functions that we would expect an IFU to serve well:

- Introduce the user to the product's features and functions
- Lead the user through specific procedures using the product
- Alert the user to take appropriate care to avoid harm and property damage
- Help the user troubleshoot problems
- Where appropriate, direct the user to perform necessary "care and feeding," meaning to clean and maintain the device properly

In the remainder of the book, we'll use the term "instructions" to refer broadly to any of the instructional materials described above.

3

Shouldn't products be usable enough on their own?

Designing intuitive products is an appropriate goal for any well-meaning product designer. But even simple or familiar products can benefit from instructions.

Consider the humble hammer. Using a hammer to drive nails shouldn't require instructions because the means of use is self-evident. Or so it would seem. There is actually a lot more to know about hammering nails that could be conveyed by instructions.

For example:

Hold the hammer at its end rather than close to the head (in other words, don't "choke up" as one would when swinging a baseball bat).

Hold the nail near the head rather than at its base. Although such a grip seems counterintuitive, it reduces the risk of hurting your fingers if you miss the nail.

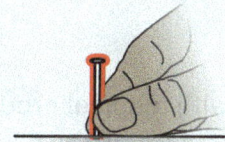

Swing the hammer by flexing your elbow rather than your wrist.

Swing the hammer so that its head is parallel to the nail at the moment of contact.

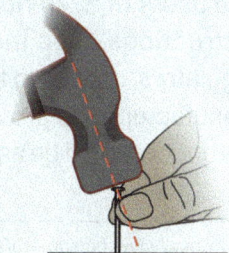

There! Did you learn something about hammering nails? Assuming the answer is yes, this is evidence that instructions can be helpful even with relatively simple and familiar products that seem intuitive to most people.

Therefore, it follows that even simple and familiar medical devices warrant instructions, which the user may choose to read or ignore. Providing instructions is not a sign of poor design. It is a sign that the manufacturer is trying to be supportive, knowing that some (if not all) product users will benefit from having access to instructions.

Let's consider the simple drug patch. If you know the patch should be worn for three days and then removed, why would you need instructions? Just put it on your skin and replace it after 72 hours.

A transdermal drug patch.

But with medical products, there are often steps for safe and effective use that are not obvious. In the case of the patch, this might include disposal instructions to fold a removed patch's sticky side onto itself to prevent the drug dispensing portion of the patch from touching somebody else, such as a child putting their hand into the trash. Also, you might want to explicitly mention the need to remove the used patch before placing a new one on the skin. Notably, there was an actual adverse event report in which a patch user suffered an overdose because she left multiple patches on her skin. For this reason, the US National Capital Poison Center provides the instruction: "Use only one patch at a time."[*]

In summary, if a simple skin patch requires instructions to be used safely and effectively, and users of even non-medical tools like hammers could benefit from more explicit instructions, then most other medical products will need them. Additionally, many regulators, including the United States (US) Food and Drug Administration (FDA), require manufacturers to provide instructions along with their medical products. Therefore, while a product should be as intuitive as possible, this does not obviate the need for instructions.

---

[*]   https://www.poison.org/articles/using-skin-patch-medicines-safely

4

Good instructions are developed as an integral part of the product

Logically, writing instructions after creating a product makes sense. After all, how can you write instructions for something that doesn't yet exist?

Indeed, you can't write complete instructions while a product is in development. But you can and should get started early for these reasons:

1. Writing preliminary instructions often reveals usability problems with the product that can be fixed more easily during early versus late stages of development

2. Preliminary instructions might be needed to support users participating in formative usability tests of the associated product, which can reveal opportunities to improve the product and instructions

3. Waiting to write instructions until a product's design is finalized can add time to the overall development schedule and delay the product's release

4. Depending on the instructional media you use, you might need a lot of lead time to produce the instructions in their final form

5. Writing preliminary instructions helps cement your strategy for teaching people how to use your product; this strategy might include other learning tools such as video or hands-on training using a simulator of the device

Beware that waiting until a product design is finished before starting the instructions might put the effort on the project's critical path and create unwelcome stress. However, as illustrated below, finishing touches on instructions will necessarily have to follow completion of the product design. See "Develop preliminary instructions" on page 61 for more details.

Identify instructional needs · Select & refine preferred design · Prepare for production

Discovery · **Concept development** · **Design engineering** · **Validation** · **Launch**

Create initial outlines & sketches · Finalize & validate content

The instructional design effort can be distributed throughout the product development process.

5

Instructions must be helpful

How do you feel when you use a software application, click on help, and the help is… not helpful? It's frustrating and it teaches you to not seek help when you need it (see *negative reinforcement loop* in "Users might have low expectations for instructions" on page 19).

Here are some analogs, other things that frustrate when they do not serve their basic purpose:

- A stapler that jams but only when you try to staple something
- An automobile infotainment system that refuses to pair with your smartphone
- A sliding screen door that derails when you move it

Now, what's the chance that one of these three sources of frustration will cause significant harm or property damage? Probably very low.

It's different for a medical device's instructions. Situations can arise in which the instructions must be helpful.

Consider another analog.

Pilots are trained to go straight to the manual (i.e., the instructions for use) to deal with an engine failure, for example. As Chesley Sullenberger piloted his stricken Airbus toward the Hudson River, co-pilot Jeffrey Skiles dutifully read emergency procedures to restart the jet engines. As it turned out, the procedures would have been useful, but the engines were too badly damaged by bird ingestion to return to full function.

US Airways Flight 1549: The Miracle on the Hudson (photo by Greg Lam Pak Ng).

Seeking guidance from instructions might be a desperate action.
Imagine a comparable emergency involving a medical device: perhaps an anesthesia machine, hemodialysis machine, or heart-lung machine. Such machines serve life-sustaining purposes and users might need information in the instructions to complete a medical procedure and protect their patient from harm.

Equally critical scenarios can play out in non-clinical environments. Consider the person who has diabetes and depends on an insulin pump to manage their blood sugar level. A device might malfunction in a way that requires quick and proper action to restore therapy, the alternative being an underdose or overdose of insulin. And even if there is no emergency, instructions must be helpful. Their only purpose, from the perspective of most users, is to help someone use a device.

Here is just a sample of how instructions can and must be helpful:

10 seconds

CLICK

Click green button and HOLD PEN firmly for 10 seconds.

Instructing users to hold their epinephrine pen in the skin for 10 seconds after pressing the inject button, or else some drug might leak out of the injection site.

3. Wash hands
Wash your hands with soap and water, then dry them completely.

Instructing users to thoroughly wash their hands before setting up their device.

6 If you see downstream air bubbles, open the resistor knob (turn counter-clockwise until ⊗ symbol aligns with arrow) to eliminate all air from the set, then close the resistor knob. Repeat as needed.

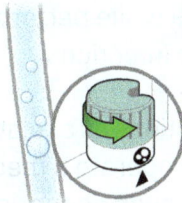

Showing users how to eliminate air from an intravenous infusion pump line that is attached to the patient.

2

Slowly remove patch
baby oil esive
a. Begin slowly peeling off adhesive patch at a low angle.

3

Remove
a. After re patch f sensc

Warning users to remove a used drug patch from their body before attaching a fresh patch.

6

Instructions can mitigate risks

Within the medical device industry, risk mitigation is paramount. Manufacturers are held to a high standard by regulatory agencies, as well as the plaintiffs' bar, to ensure that products function properly and, if they fail, do so in ways that do not harm someone or cause property damage.

When products do fail, the consequences can be devastating. People might be injured or killed. People who use medical devices to treat others feel intense guilt and might even face prosecution. Medical device manufacturers might be sued and consequently go out of business.

This book focuses on producing instructions that help to prevent use errors, which are also called mistakes, and are a type of failure. In fact, medical device manufacturers track use errors in the same manner as other device failures, such as a short circuit, in the process of mitigating risk. For example, many manufacturers perform a failure mode and effects analysis of foreseeable use errors. This allows manufacturers to figure out how to (1) reduce the chance that the use error will occur and (2) reduce the severity of harm or damage that might result. In other words, they engage in risk mitigation, and this is where instructions are in play.

The best way to mitigate use-related risk is to eliminate the chance of a use error occurring. For example, if an injection needle should only be used once, the injector may be designed to enable single use and be impossible to use again, making it inherently safe in at least one important regard. In lieu of an inherently safe solution, it may be possible to add a protective feature that guards against use error, such as a clear cover over a critical control to avoid accidental actuation. Next in the hierarchy of risk mitigation is adding a warning that encourages safe behavior (e.g., wash hands before performing an injection) or discourages unsafe behavior (e.g., do not recap the used needle).

Finally and logically, instruction contents can help mitigate risk. They can keep users on-track versus veering dangerously off-track while performing a critical procedure, such as cleaning contact lenses to prevent eye infections.

However, it is good practice to consider instructions as a relatively weak risk mitigation for one particularly strong reason. A user can operate a medical device without even looking at the accompanying instructions. In such scenarios, the instructions never have a chance to help the user avoid mistakes. There is also the compromising scenario wherein the user might want to read the instructions but cannot find them; a common occurrence when a medical device has been in use for a long time and the instructions have been lost.

Accordingly, a good way to think about instructions is that they can sometimes serve as a helpful risk mitigation and therefore warrant an investment in good design. However, they should be considered a complement to other risk mitigations rather than the primary risk mitigation.

7

Instructions must meet regulatory requirements

In most countries, governmental organizations regulate the medical product market. The associated regulations affect medical device instructions as well.

Here are just a few of the regulatory agencies that uphold technical requirements for medical products and control market access in their respective countries:

🇧🇷	Brazil	Brazilian Health Regulatory Agency (Anvisa)
🇦🇺	Australia	Therapeutic Goods Administration
🇨🇦	Canada	Health Canada
🇨🇳	China	National Medical Product Administration
🇪🇺	European Union	European Medical Agency
🇯🇵	Japan	Pharmaceuticals and Medical Devices Agency
🇸🇦	Saudi Arabia	Saudi Food and Drug Authority
🇬🇧	United Kingdom	Medicines and Healthcare Products Regulatory Agency
🇺🇸	United States of America	US Food and Drug Administration

Not all regulators have specific requirements for instructions. Some regulators simply require manufacturers to provide instructions together with their products, while others have more specific requirements for the type and format of instructions included with medical products. For simplicity's sake, in this section we'll summarize the regulations set forth by two of the most prominent regulatory bodies: the US Food & Drug Administration (FDA) and the European Union's European Medical Agency (EMA).

FDA regulations

The legal basis for the FDA's regulation of all products under its provision (e.g., cosmetics, foods, medicine, etc.) is defined in a series of documents titled Code of Federal Regulations Title 21, or 21 CFR for short.

Medical devices are covered in parts 800 to 899 of 21 CFR.[*] Regulations for medical device instructional materials are defined in part 801, which addresses labeling — a broad category that the FDA defines as "all written, printed, or graphic matter accompanying an article at any time while such article is in interstate commerce or held for sale after shipment or delivery in interstate commerce."[†]

Part 801 provides a series of fundamental requirements for medical product labeling, many of which can be relevant to instructions as well, such as:

- Labeling must not be false or misleading (§801.6)

- Labeling must contain a statement of identity that includes the common or usual name of the device, followed by a statement describing its intended actions (§801.61)

- Labeling must declare the quantity of the device's contents (e.g., components and accessories) (§801.62)

- If symbols are used in the labeling, they must be explained in a glossary that is included in the labeling and provides a clear explanation of each symbol used (§801.15(e))

The regulations also include requirements specific to instructions, which distinguish between instructions intended for laypeople and those intended for healthcare professionals.

Section 801.5 states that devices intended for use by laypeople must be accompanied by "adequate directions for use," meaning that they enable non-healthcare professionals to operate the associated device safely for its intended purpose. The section also identifies the following content as necessary for instructions to be adequate:

- The health or medical conditions associated with the device (e.g., diabetes)

- The device's intended use (e.g., measuring blood glucose levels)

- The device's purpose (e.g., monitoring blood glucose levels as part of diabetes management)

---

[*] Available on FDA's website at https://www.ecfr.gov/current/title-21/chapter-I/subchapter-H
[†] FDA CFR - Code of Federal Regulations Title 21 section 1.3(a)

- Details regarding the timing of the device's use (e.g., frequency, duration, specific times of day, etc.)

- Quantity of dose, including different amounts for different ages and physical conditions, if applicable

- The route or method of administration or application (i.e., how to use the device)

- The preparation for use (i.e., how to set up the device)

Section 801.109 states that prescription devices intended for use by healthcare professionals are exempt from requiring adequate directions for use. At first glance, this might be interpreted as suggesting that medical devices for healthcare professionals don't require instructions. However, a closer reading reveals that the section uses the term "information for use" to refer to instructions intended for healthcare professionals. The section also states that such instructions should include the following details:

- A statement indicating that the device is intended for prescription use only (e.g., Rx only") or the caution statement "Caution: Federal law restricts this device to sale by or on the order of a [practitioner's title]"

- The device's indications for use (i.e., what the device is used for)

- The device's effects (i.e., results the device is expected to produce when it's used as intended)

- Methods and routes for using the device (i.e., how and where on the patient's body the device is intended to be used)

- Frequency and duration of device use

- Risks and hazards associated with the device's use

- Contraindications for using the device

- Possible side effects associated with the device's use

- Precautions to be taken when using the device

- The date on which the instructions were created or last updated

Note that per §801.19(c), manufacturers may omit the above details from instructions "... if, but only if, the article is a device for which directions, hazards, warnings, and other information are commonly known to practitioners licensed by law to use the device. Upon written request, stating reasonable grounds therefor, the Commissioner will offer an opinion on a proposal to omit such information from the dispensing package under this proviso."

Also note that Section 801 includes subsections pertaining to specific types of medical devices, including in vitro diagnostic products, dentures, hearing aids, and medical devices containing natural rubber. As we describe in "Review regulatory guidance" on page 50, it's a good practice to review the current version of 21 CFR, as well as the FDA's latest guidance for instructions, when embarking on a new instruction development effort.

EU MDR requirements

The legal basis for the regulation of medical products intended for release in the European Union (EU) is defined in the EU Medical Device Regulation (MDR), specifically Regulation (EU) 2017/745, which is the latest version at the time of this writing. The MDR mandates that instructions for use (IFUs) must accompany medical devices and provides specific requirements to ensure they facilitate the safe and effective use of their accompanying devices.

The MDR outlines the information these instructions must contain. The required information generally mirrors that required by the FDA, including:

- The product's intended purpose and user population
- Contraindications and side effects
- Required product preparation (i.e., setup)
- Step-by-step usage instructions
- Information on cleaning, maintenance, and disposal of the product
- Applicable warnings or cautions

In addition to these general requirements, the MDR provides guidelines for specific products. For instance, it mandates that implanted products, such as pacemakers, come with an "implant card" that identifies the product and provides instructions for safe use, warnings, and details about the implant's expected lifespan.

The MDR also includes recommendations for electronic instructions for use (eIFUs). Notably, it only allows certain devices, such as fixed and installed medical devices or those with built-in screens intended exclusively for use by healthcare professionals, to have only electronic instructions. When opting for eIFUs, manufacturers must provide sufficient evidence that the eIFU does not introduce additional risk to product use and must clearly label where users can find the instructions on the product itself, such as with a prominently affixed label.

8

Instructions should harmonize with the product

Instructions and the associated medical product should have a common look and feel (i.e., harmonize).

A good first step toward creating a harmonized appearance that reinforces the connection between the instructions and the product is to follow the manufacturer's brand guidelines. Such guidelines usually prescribe the use of color, typography, and terminology that might be found across products, instructions, and promotional material.

For example, the instructions accompanying the medical devices shown below could use the primary product colors as the background to major headings and use the companies' preferred typefaces.

PulseRevive
QUICK START GUIDE

Step 1
Push "Power" button

3 Attach the needle

Such continuity of branding is likely to be pleasing to users, particularly as compared to cases where the instructions and associated medical product are visually unrelated. However, instruction designers need to prioritize usability and document design principles over branding goals. Otherwise, certain visual features might distract or confuse, thereby leading to use errors.

Here are some additional guidelines for creating effective harmony between the product and its instructions:

- Implement branding elements in a relatively subtle, rather than garish, manner so that users can focus on key content.

- Make sure there is sufficient contrast between backgrounds and overlaying text to ensure legibility.

- Reserve the use of certain colors to convey special meanings, such as using red to denote a warning.

- Use language and symbols that are consistent with those used in the product's user interface. To ensure consistency, consider developing a style guide that documents standard terms and symbols for use in the product and its accompanying materials.

Two sample designs that show different ways of applying the same branded color palette.

- Consider using a brand-specific font for major headings, which presumably will be larger than the body text. However, if the font is highly stylized and ill-suited for reading sentences and paragraphs, use a more suitable font for the body text.

Finally, the instructions and the product should use the same terms, which may be specified in a style guide. The following table is an example of a terminology guide that medical product manufacturers might include in a style guide to suggest standard terms in lieu of other options.

Standard terms	Terms to avoid
Blood tubing set	Dialysis set, tube set, tubes, set
Pen-injector	Injector, pen, delivery device
Start pump	Activate pump, turn on pump
Diluent	Saline, reconstitution fluid
Electrode pad	Pad, electrode, dispersive pad, conductor

A sample terminology table from a hypothetical style guide.

9

Users might have low expectations for instructions

There is something called a negative reinforcement loop. The concept applies to many things, including older cars. Consider the person who owns a car that already has 125,000 miles on it. They might not spend a lot on maintenance because they don't expect the car to last much longer. Consequently, the car is poorly maintained and ultimately breaks down more often, thus reinforcing the negative impression that the car is unworthy of further maintenance.

An earlier generation of Volvo cars were the beneficiary of a positive reinforcement loop. People thought their Volvos were built to last a long time, so people reportedly spent more on maintenance. As a result, Volvos lasted a long time, and not just because they were built to last. Well, at least that's the story that remains in circulation.

So, how does this all relate to instructions?

Some medical device manufacturers seem adamant that people do not read instructions. This leads them to conclude that an instructional document is obligatory but not worth a large investment. This results in objectively lousy instructions, which users consider useless and ignore. There's your negative reinforcement loop.

Here are some major telltales that a manufacturer had disdain for their medical device's instructions:

- The instructions seem like they were written by engineers for engineers. They do not "speak" to the user in a way that will be fully comprehensible.

- The instructions appear to be cheaply produced, evidenced in shortcomings such as thin, see-through paper and black and white printing.

- The content does not ascribe to good design practices that help with information navigation, ensure legibility, improve readability, and enable rapid information acquisition.

Decades of poor instructions took their toll. Medical device users expected instructions to be poor and treated them as a last resort, preferring to ask people for help or take a trial-and-error approach to device use. Not good.

Fortunately, some manufacturers seem committed to producing excellent instructions. In our opinion, excellent instructions are a strong indication of excellence in every aspect of product design and engineering as well as the user experience.

Keep in mind the rule of thumb: "Instructions will help some people, some of the time, to some extent." Kudos to the manufacturer who understands this and cares about their users enough to meet the need for excellent instructions.

Note: Volvo is still in the car business, albeit under new ownership, and they produce what the public perceives to be nice, long-lasting cars.

DON'T FORGET!
Instructions will help
SOME people
SOME of the time
to **SOME** extent

10
Users can be delighted by good instructions

To delight someone is to give them great pleasure.

Advertisements for traveling circuses promise that adults and children alike will have a delightful time by providing a unique experience with varied entertainment, which could include anything from a surprising clown to a fearsome animal tamer.

Vintage circus advertisements.

Products can delight users as well by providing deeply satisfying and pleasurable experiences. You probably have favorite items that might include your smartphone, backpack, office chair, non-stick frying pan, or smart thermostat. Similarly, medical products such as glucose meters or walkers can delight users by meeting their needs and then some.

Realistically, instructions for using a medical product cannot compete with the circus, let alone a smartphone, for entertainment value. But they can delight in their own way. When someone is apprehensive about using a medical product correctly and the instructions make things easy, they might very well be delighted. Maybe this is more akin to coming away from a dreaded dental procedure that went smoothly, now delightfully painless after suffering from a toothache.

Here is another example. Pharmacies sell myriad test kits, including for COVID, HIV, and various recreational drugs. At the time of this writing, all three book authors had recently tested for COVID using rapid test kits made by different manufacturers. One out of three was pleased by the quality of the instructions that came with the test kit, the other two not so much. In the one case, delight sprang from an initial concern that testing was complicated and prone to use error, but it turned out to be straightforward thanks to the instructions' clear illustrations and plainly worded prompts. The other two were frustrated by their kits' small text, vague instructions, and illegible illustrations.

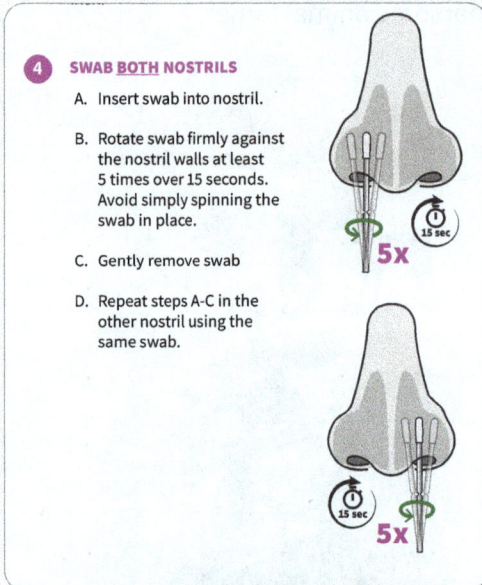

Examples of instructions from two different rapid COVID tests.

This book covers many aspects of instruction design and writing that can contribute to a delightful end product. As a preview, here are what we consider the five most significant factors in creating delightful instructions (noting that neglecting other factors can undermine the final result):

1. Speak directly to users with plain language
2. Use illustrations liberally
3. Make procedural steps stand out
4. Include content that helps users avoid potential errors and not just recover from them
5. Arrange information in a clear way that guides users through the instructions
6. Use color in an informative way

11
Good instructions can generate goodwill

A product that satisfies users can enhance the manufacturer's reputation. In other words, it can generate goodwill that might lead to loyal, repeat customers.

So, can instructions generate goodwill in the same manner as a well-designed product? Of course they can. This is because instructions are a highly visible part of the product's user interface.

Customer goodwill is an intangible asset that businesses obtain through providing consistent, high-quality customer service. It accounts for non-quantifiable returns that are difficult to measure and categorize, like customer loyalty, brand reputation, and customer value.[*]

In many ways, instructions represent a company's voice that speaks directly to its customers. In place of a spokesperson or trainer explaining how a product works, the instructions do it. Moreover, they speak to a customer at pivotal moments, such as when using the product for the first time and when running into trouble using it.

As consumers, the last thing we want is to open an instruction manual and not understand how to begin using our new product. That's the last thing businesses want as well; they want their new users to set up and begin using their new purchase quickly and easily.[†]

When instructions perform well, users are more likely to appreciate the company that created them. Similarly, if a user struggles to use a product and the instructions help overcome their challenges, their appreciation rises for the instructions and the company. Consider it a win.

We expect that the following examples of real-world medical product instructions generate goodwill toward the manufacturers.

---

[*] Clint Fontanella, What Is Customer Goodwill? [Definition + Examples]. April 27, 2020.
    https://blog.hubspot.com/service/customer-goodwill
[†] Flori Needle. 4 Enjoyable & Effective Instruction Manuals. October 6, 2021.
    https://blog.hubspot.com/service/effective-owners-manuals.

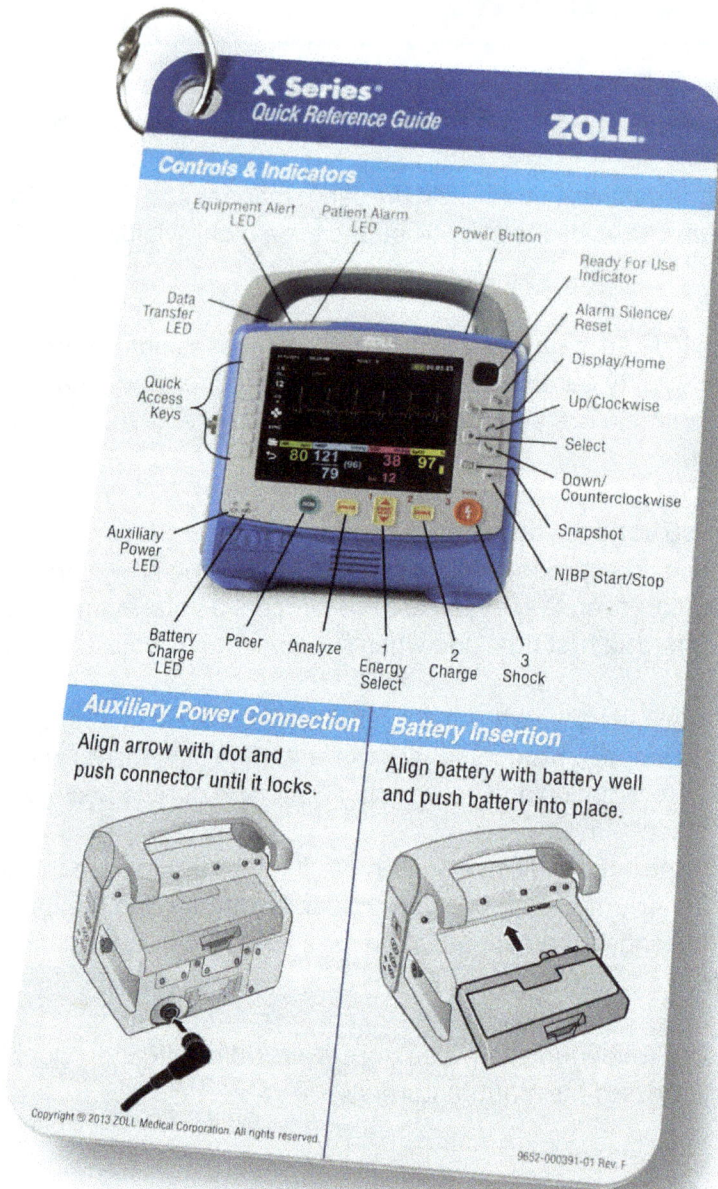

A quick reference guide for a defibrillator.
(Image courtesy of ZOLL Medical Corporation)

EASYGRIP Flo-41
PRECISION MIS DELIVERY SYSTEM

QUICK REFERENCE GUIDE

1 Initial & Repeat Product Set-up and Application
Inside the sterile field

- **a)** Attach the FLOSEAL Matrix syringe to the EASYGRIP FLO-41 fill port. **b)** Fill the reservoir completely with FLOSEAL Matrix. **Syringe may be removed if desired.**

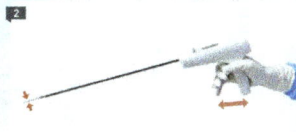

- Prime cannula by repeatedly pulling the trigger (approximately 2-3 pulls) until FLOSEAL Matrix is visible in the clear malleable tip. The EASYGRIP FLO-41 Applicator is now ready to use.

- Pull the trigger of the EASYGRIP FLO-41 to apply the FLOSEAL Matrix, per its instructions. Multiple pulls may be required to apply desired amount of FLOSEAL Matrix.

  If additional FLOSEAL Matrix is required during the same surgical procedure, repeat steps 1&3. Priming is not required for repeated loading.

2 Residual Product Application
Inside the sterile field

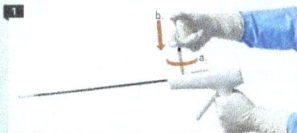

- If application of residual FLOSEAL Matrix is needed, **fill the provided 1.5 mL syringe with non-heparinised saline. a)** Attach saline syringe to EASYGRIP FLO-41 and **b) fill the reservoir with only 1.5 mL of non-heparinised saline.**

- Pull the trigger of the EASYGRIP FLO-41 Applicator System to apply residual FLOSEAL Matrix at the site of the bleed.

- If, after expelling all residual product from cannula, additional FLOSEAL Matrix application is required, the cannula must be purged of saline. **Remove saline from the cannula and prime the device for additional use by repeating steps 1 & 2 above.**

Please see Indication and Detailed Important Risk Information on the reverse side.

Baxter

Excerpt from a quick reference guide for a surgical instrument.
(Image courtesy of Vantive Health LLC)

TAKING YOUR DOSE

① Open

Set the inhaler
on a flat surface.

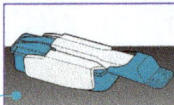

Open the inhaler by
lifting the mouthpiece
to an upright position.

② Load

Place the cartridge
in the inhaler.

Bring the mouthpiece
down to close the inhaler.

You should feel a *snap* when
the inhaler is closed.

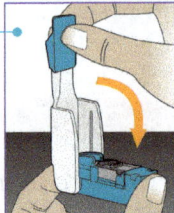

③ Inhale

Remember to read all of the instructions below before taking your dose.
For more information, see the TYVASO DPI Instructions for Use manual.

Remove the blue mouthpiece
cover, pick up the inhaler
while keeping it level, and fully
exhale away from the inhaler
before taking your dose.

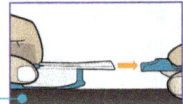

Keep your head level, seal your
lips around the mouthpiece, and
tilt the inhaler slightly downward
to prevent your tongue from
blocking the powder.

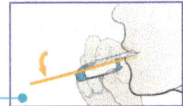

Breathe in deeply through
the inhaler.

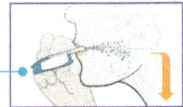

Take the inhaler out of your
mouth, hold your breath for as
long as you comfortably can,
and blow out before continuing
to breathe normally.

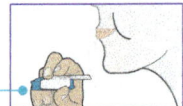

If your dose requires more than 1 cartridge,
repeat the previous steps for each cartridge.

TYVASO DPI
(treprostinil) INHALATION POWDER

Excerpt from a quick reference guide for a dry powder inhaler.
(© 2022 United Therapeutics Corporation. All rights reserved. Used with
permission. TYVASO DPI is a trademark of United Therapeutics Corporation)

Start Guide

ZIO® BY iRHYTHM

1. Position Monitor

Enroll the Patient
Online: www.ZioSuite.com
Phone: 1.888.693.2401

Write in Patient Name, Start Date, and Prescribed Wear Time on cover of Symptom Log

Have the patient stand up with their arms relaxed by their sides. If they're unable to stand, have them sit in a chair with their arms relaxed

Determine placement without removing backing:
- Find the area where the monitor will be placed on the patient's upper left chest. It is about one finger width below the left collarbone
- Zio monitor will be placed diagonally on the chest after the skin is prepared

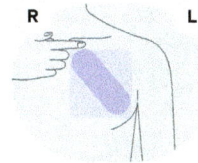

R   L

2. Prep Skin

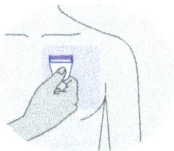

Shave area: use the provided razor to shave the entire placement area (all genders, including those with no visible hair)

Dry area: make sure the placement area is fully dry

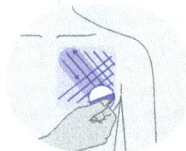

Exfoliate: use the rough side of the provided exfoliator to gently exfoliate the entire prep area. Complete 40 strokes in total: up and down, side to side, and both diagonal directions

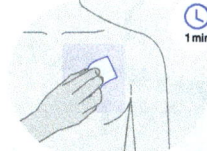

1 min

Clean area: use the provided alcohol wipes to clean the area thoroughly

Allow the skin to dry for at least one minute before continuing to the next step

3. Apply Monitor

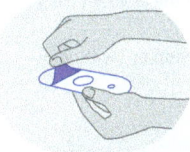

- Identify the area one finger width below the patient's collarbone
- Carefully remove the clear plastic backings from the back of the monitor to expose the adhesive

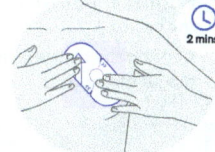

2 mins
- Place the Zio monitor diagonally on the prepared skin area, one finger width below the left collarbone. Make sure the arrow is pointing upward
- Massage the paper tabs firmly for two minutes

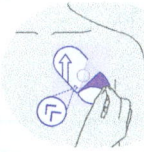

2 mins
- Find the double arrows on the paper tabs located above and below the center of the Zio monitor. Starting with one tab, peel in the direction of the double arrows
- Using your other hand, ensure the adhesive wing does not lift
- Repeat these steps to remove the remaining tab
- Massage the adhesive wings firmly onto the skin for another two minutes

4. Activate Monitor

- Press and release the button at the center of the monitor. You'll notice a green light flash briefly
- Have your patient keep the Symptom Log, Patient Guide, and Box

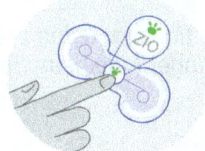

A start guide for a wearable cardiac monitor.
(Used with permission by iRhythm Technologies)

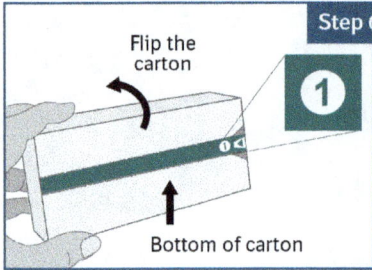

Step 6a	**Get to know carton.**

Flip the carton

1

Bottom of carton

Sit at the table for the following steps:

Rotate the carton to view the green perforated peel-off strip.

Find the number **1** .

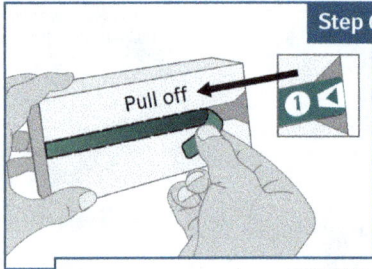

Step 6b	**Unseal and position carton.**

Pull off

1 ◀

Lift the **1** and pull the peel-off strip completely off the carton.

Dispose of the strip in the general waste.

Flip the carton over towards the table so the label is facing the ceiling.

Confirm the carton is in the correct position by locating the number ② .

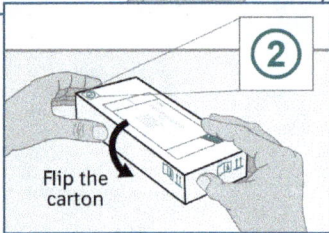

②

Flip the carton

Excerpt from instructions for using pre-filled syringe.
(Images courtesy of Boehringer Ingelheim Pharma GmbH & Co. KG)

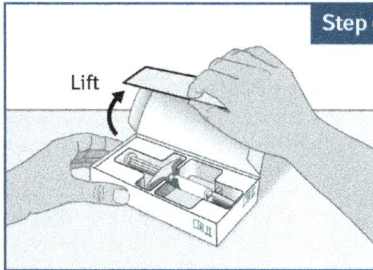

Step 6c Open the carton by lifting the lid up.

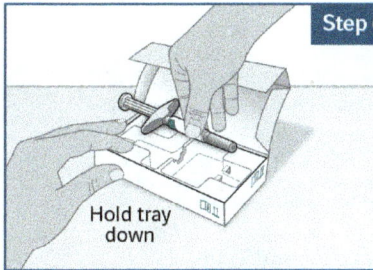

Lift

Step 6d Remove prefilled syringe.

Hold the tray down and grab the prefilled syringe by the barrel to remove it from the carton.

Place the prefilled syringe on a clean, dry and flat table in a quiet and well-lit area near the already gathered supplies.

- **DO NOT** dispose of the empty carton. Keep it for the next trial site visit.

Hold tray down

Excerpt from instructions for using pre-filled syringe.
(Images courtesy of Boehringer Ingelheim Pharma GmbH & Co. KG)

GETTING STARTED GUIDE

Nellcor™ bedside respiratory patient monitoring system, PM1000N

UNBOXING AND SETTING DEFAULTS

1 Unbox the monitor

Contents:

- Nellcor™ bedside respiratory patient monitoring system, PM1000N
- DOC-10 interface cable
- Operator's Manual and/or CD
- Hospital-grade power cord
- Internal lithium-ion battery

2 Install the battery

- Using a T15 X 3-inch TORX screwdriver, remove the battery cover located on the right side of the rear panel.
- Insert the battery with the white tab facing out and on top, so the battery connection terminal on the opposite side is facing down.
- Replace the battery door.

3 Connect to an AC power source

- Plug the female connector end of the power cord into the power connector on the rear of the monitoring system (opposite of the battery door).
- Plug the male connector of the power cord into a properly grounded AC outlet.
- Verify the monitor AC power indicator lights (green light next to the ∿ symbol).

4 Connect DOC-10 interface cable

- Depress the buttons on both sides of the cable connector as you insert it into the monitor sensor port.
- Release the buttons after the cable connector is firmly in the socket.

5 Power the monitor on

- Press the **Power** key.
- The monitor will perform a self-test followed by three ascending tones and a one-second beep.
- Patient monitoring screen will appear after self-test successfully completes.

6 Setting institutional defaults

- Reboot the monitor to enter the service mode. Use the **Power** key to turn the monitor off and back on.
- Press the wrench icon when it appears on the self-test screen (right after the three ascending tones).
- Enter the service mode password: 62907.
- Monitor is now in **Service** mode and ready to set defaults.

See page two for important additional details on setting institutional defaults.

7 Set Date and Time

- Using the touch screen, press **Menu**.
- Press **Monitoring Settings**.
- Press **Time and Date**.
- Using the plus and minus keys, navigate through the **Time and Date** settings and choose the format you prefer to see the date.
- Select **Save Changes**.

Medtronic

Further, Together

Excerpt from Nellcor™ PM1000N respiratory rate monitor quick reference guide. (©2024 Medtronic. All rights reserved. Used with the permission of Medtronic.)

Medtronic

Puritan Bennett™ 980 ventilator

Quick reference guide: Touch screen

Touch screen icons

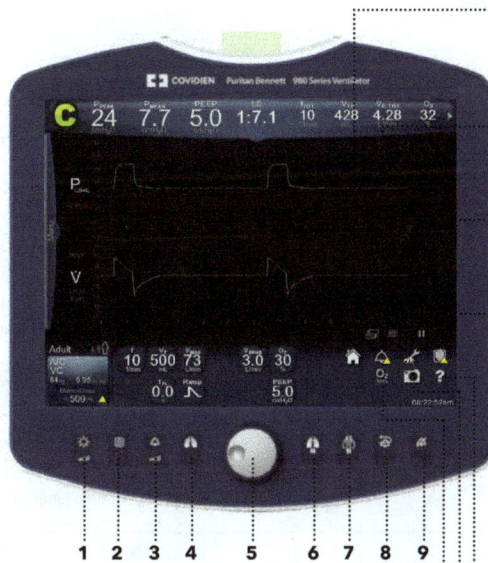

Bezel keys

1. Display brightness key
2. Display lock key
3. Alarm volume key
4. Manual inspiration key
5. Rotary encoder dial
6. Inspiratory pause key
7. Expiratory pause key
8. Alarm reset key
9. Audio pause key

Home
Touch icon to close all open dialogs and return to the main screen – displayed here.

Alarm
Touch icon to display the alarm settings screen.

Configure
Touch icon to display the configure screen.

Logs
Touch icon to display the logs screen which contains:
• Tabs for alarms
• Settings
• Patient data
• Diagnostics
• EST/SST status
• General event
• Service logs

Help
Drag icon to the item in question and release – a tooltip will appear describing the item's function.

Screen Capture
Touch icon to capture and store the image displayed on screen.

Elevate O$_2$
Touch icon to configure and increase the oxygen concentration for 2 minutes.

This guide is provided as a convenient companion document to the operator's manual. It's not intended to replace the operator's manual, which should always be available while using the ventilator. It's important to familiarize yourself with all information in the operator's manual relevant to your institution's use of the ventilator, including on-screen help, instructions, warnings, and cautions.

Excerpt from Puritan Bennett™ 980 ventilator quick reference guide.
(©2024 Medtronic. All rights reserved. Used with the permission of Medtronic.)

Torso-One and Torso

Quick Start Guide for Bridge

Getting Started

Step 1   🔋 Charge the Bridge

Plug the power supply into the USB-C port on the side of the Bridge

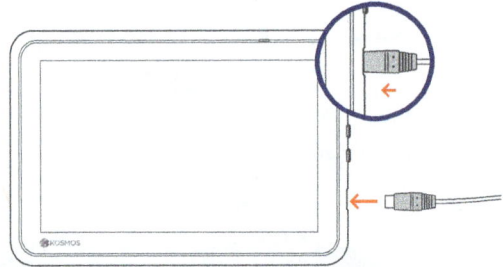

Step 2   📶 Connect to Wi-Fi (Optional)

From the Kosmos Homescreen, tap
Settings > Admin > Wi-Fi

USB-C port for charging and Lexsa connector

Step 3   Connect Torso-One/Torso

Plug the Kosmos Torso-One or Torso connector into the port under the Kosmos Bridge Handle.

Note: To register your transducer and licensed features for the first time, the probe must be connected to the device and your device must be connected to the internet. This step may take a few minutes.

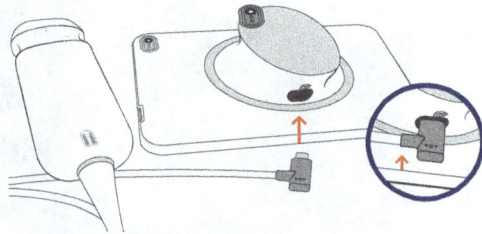

USB-C port for Torso-One/Torso connector

Step 4   Transducer Element Check

An automatic Transducer Elements Test is initiated every time a Kosmos probe is connected to a device.

With the completion of a successful test, users can begin scanning.

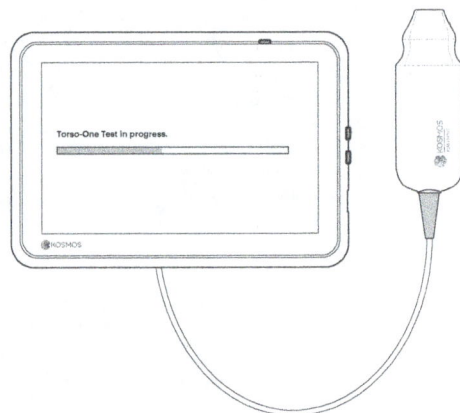

Torso-One Test In progress.

2   Torso-One and Torso for Bridge

Excerpt from an ultrasound system quick start guide.
(Image and content courtesy of EchoNous, Inc.)

8.2 Switch on the tonometer

⚠️ **PRECAUTION!** The tonometer switches off the display when it has not detected any movement for 15 seconds. The tonometer switches off automatically if it has not been used for 3 minutes.

Make sure that the date and time shown on the display are correct. If they are incorrect, update them from the tonometer's settings or by connecting the tonometer to the iCare PATIENT2 application or to the iCare EXPORT software.

Press down ▶ until you hear a beep. The text "Start" is shown on the display.

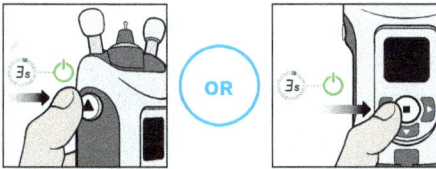

Alternatively, press down ⊙ until you hear a beep. Then press ⊙ again to enter the measurement mode. The text "Start" is shown on the display.

8.3 Find the correct measurement position

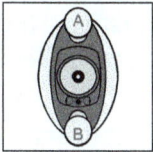

The forehead support A rests on your forehead and the cheek support B rests on your cheek.

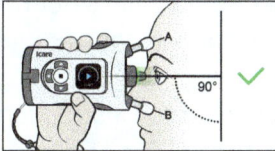

Look straight ahead and the tonometer is at a 90-degree angle to your face. The probe is about 5 mm (3/16 inches) from your eye and points perpendicularly to the center of your eye.

NOTE: The tonometer measuring button should point upwards.

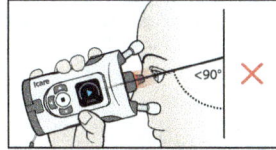

If you see a red light in the probe base, the tonometer is tilted too much downwards. You should straighten your posture and lift your chin.

8.4 Adjust the supports and position the tonometer

⚠️ **WARNING!** Only probes are intended for contacting the eye. Avoid touching the eye with other parts of the tonometer. Do not push the tonometer into the eye.

⚠️ **WARNING!** Shorten the cheek and forehead supports of the tonometer only slightly at a time to prevent the tonometer from getting too close to your eye.

⚠️ **PRECAUTION!** Eye detection is based on the difference of infrared reflections received from the transmitters: the nose side reflects more than the temple side. If the transmitters become dirty, the recognition may be interfered.

⚠️ **PRECAUTION!** Do not cover the eye recognition transmitters or sensor during the measurement, for example with your fingers. Keep your hand, hair, and objects such as pillows away from the temple side of your eye, as they produce an infrared reflection that causes an error.

1. **Before the measurement, adjust the forehead and cheek supports to the correct length. Start with the supports at the maximum length.**

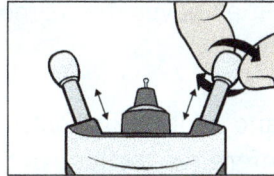

You can take the measurement sitting, standing, or lying down (supine position).

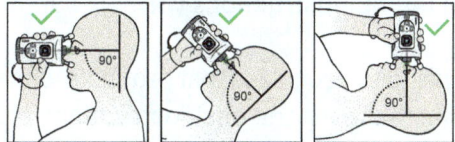

You can hold the device with one or both hands.

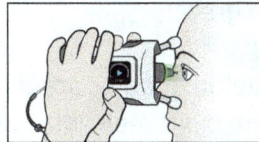

17

18

Excerpts from instructions for using a handheld tonometer.
(Images and text courtesy of Icare Finland Oy)

12
Satisfying users

Lousy instructions are disappointing, if not infuriating.

For example, a "Getting Started" guide that is overly simplistic and disorganized insults a user's intelligence, wastes their time, and could lead to mistakes that damage the product or even increase the risk of injury to the user.

Sub-par instructions can also make a user angry with the product manufacturer. It can really damage customer goodwill.

Now consider an exemplary "Getting Started" guide. Sure, despite its fine design, a user might completely ignore it, that is, until they face some sort of difficulty and need help. On the other hand, a user might use the instructions for their intended purpose — learning about the product and following its instructions through the initial steps of obtaining a blood pressure reading.

In contrast to sub-par instructions, users might sense that a good one complements their natural abilities, saves them time, and protects against detrimental mistakes (i.e., use errors).

Well-designed and informative instructions, however, do not guarantee that users will appropriately reference the instructional materials at the correct time and with the necessary attention, as discussed earlier in this introduction. So, here again is our rule of thumb: instructions will help some people, some of the time, to some extent.

This rule recognizes that products with superb user interfaces might not need instructions and that even the best instructions are unlikely to be universally helpful to all people in every way. Yet, instructions are worth the investment to make them reasonably good and perhaps even great.

The cost of producing good or great instructions is almost certain to be less than the cost of dealing with customers who cannot get a product to work properly due to lousy instructions. This benefit-to-cost assessment does not even account for the cost associated with personal injury and property damage that could arise from use errors. Regarding this last point, you can think of good instructions as an important use-related risk mitigation, as well as a business risk mitigation.

13

Can't AI make instructions for my product?

Artificial intelligence is conquering the world. Don't worry — it's not coming for you (just yet). But, might it eliminate your need to develop instructions?

During the writing of this book, the field of AI has seen unprecedented and rapid advancement. The release of so-called "generative" AI tools, which can create new data based on learnings from existing data, have dramatically changed expectations for what AI can do. Such tools include:

- Large language models (LLMs), such as Anthropic's Claude and OpenAI's ChatGPT, which can analyze and generate text and seem to perform some level of human-like reasoning

- Image synthesis models, such as Midjourney and OpenAI's DALL-E, which can generate images based on user inputs

As instructional designers, we have been equally excited by and concerned about the potential impact of such tools.

Let's begin by discussing some of our concerns. First, the existential one: will AI eliminate the need for human designers, illustrators, and writers to develop instructional materials? Perhaps eventually, but probably not soon (as of 2024).

Sure, you could prompt ChatGPT to develop written content for an instructional document, and instruct Midjourney to produce illustrations, and they could produce something that, at first glance, might resemble decent instructional content. But, upon closer inspection, it is likely to appear as a false facsimile of effective instructional content. More importantly, it is likely to be inaccurate, as any content generated by AI is informed by the data on which it was trained, rather than the particularities of your product. In addition, generative AI has a tendency to "hallucinate," meaning to produce false information or unrealistic images.

There are also ethical and legal concerns associated with the use of generative AI. Generative AI tools are trained on massive amounts of data, including instructional materials and illustrations, much of which is obtained without the rights' holders consent. Even though generative AI systems are intended to produce new content, the potential exists for them to create content that plagiarizes from the data on which the systems were trained.

Despite these shortcomings and concerns, there are certainly productive and helpful uses for AI in instructional design development. As a general principle, it's helpful to think of generative AI not as a substitute for you, but as an assistant that can complement your skills. Potential uses for generative AI tools include:

Potential use	Sample prompt
Identify potential content to add to an outline	*Below is an outline for a [product type] [document type]. Suggest additional content or sections that would enhance the completeness and usefulness of the [document type]. [outline]*
Develop alternative conceptual models or outlines based on a draft outline that you've created	*Below is an outline for a [product type] [document type]. Generate 3 alternative conceptual models or outlines that could provide different but effective structures for this [document type]. [outline]*
Identify points of confusion or misunderstanding in a draft text	*Review the following draft text for a [product type] [document type] and identify any points of confusion or areas where users might misunderstand the instructions: [draft text]*
Identify potential misinterpretations of a given text	*What are all the possible ways that a reader might misinterpret the following text: [draft text]*
Identify opportunities to shorten a text	*Make the following paragraph as concise as possible without losing information: [draft text]*
Identify ways to make text more suitable for a particular reading level	*Adjust the language and complexity of the following text to make it suitable for a [reading level] audience: [draft text]*
Proofread draft texts for grammatical errors and typos	*Proofread the following draft text for a [product type] [document type] and list all grammatical errors, typos, or awkward phrasing: [draft text]*
Adapt text instructions to a draft script for an instructional video or for audio-based instructions	*Adapt the following written instructions for a [product type] into a clear and engaging script suitable for a 3-5 minute instructional video: [written instructions]*

Potential use	Sample prompt
Generate translated text for testing layout options	*Translate the following text for a [product type] [document type] into [target language] and ensure the translation maintains the original meaning: [text]*
Check the consistency of terminology and style across a document	*Review the following [product type] [document type] for consistency in terminology and style. List any items or concepts that are described using multiple terms: [draft document]*
Identify content that might be insensitive to cultural norms	*Review the following text and identify any ways in which it might unintentionally offend or upset readers: [draft text]*
Brainstorm ideas for icons, symbols, or illustrations	*I'm trying to create an icon that visualizes a reminder to take a medication. Suggest 10 different ways that I could show this visually.*
Create placeholder illustrations for an early mockup	*Generate an image of a hand placing a cap on a syringe. Use a clean, simple visual style.*

Importantly, be critical when reviewing answers or content provided by generative AI. This is particularly true of LLMs, which tend to respond with an authoritative tone and give a misleading impression that its answers are true. Use your judgment to determine whether the LLM's suggestions would improve or weaken your instructions. That said, even incorrect suggestions might inspire you to think of novel improvements to your instructions.

Also, be wary of considering LLMs as a proxy for users or as a substitute for usability testing. Although an LLM can pretend to be a particular type of user and judge a draft instruction accordingly, its interpretation of the instruction is likely to be quite different than that of a typical user.

Finally, you might want to consider the possibility of using an LLM as a medium for providing instructions to users. Using a tool like OpenAI's Custom GPT, you could train an LLM on an extended set of instructions for your product and have the LLM serve as a chatbot-based instructor with whom users can interact. That said, as of this book's publication date, medical product regulators have not yet indicated whether this would be an acceptable format for delivering instructions.

CHAPTER TWO
DEVELOPMENT STEPS

14

Determine user needs

Much has been written about conducting user research to determine user needs. In fact, there are several disciplines focused on the task, including the practices of human factors engineering, ethnography, and market research. Each discipline has its way of getting to the heart of what users need in a product. Choose your preferred flavor, so to speak, and use the research terminology and methods you prefer.

In our view, determining user needs always comes down to the following: ask good questions and listen carefully. You might also watch people interacting with products — usually in a non-obtrusive way — to identify unspoken but observable needs.

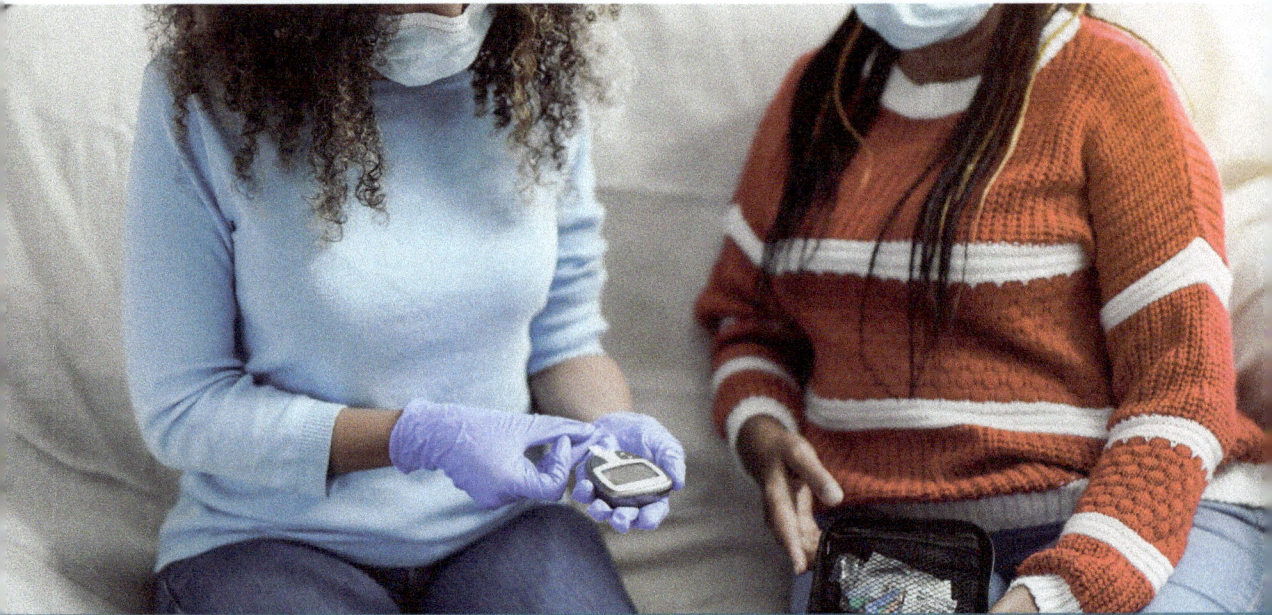

Researchers can learn a lot about users' needs by watching them interact with a product.

The principle of keeping things simple sometimes favors conducting user interviews, although there are other approaches to gathering user feedback. One-on-one interviews are very productive but also time consuming if you want to hear many voices. A group interview or so-called focus group speeds things up by including more people (typically 5-8) at once but limits how much any one person can share about their needs. You can also conduct surveys, which are analogous to impersonal interviews that can produce an even larger number of responses but often without the more nuanced findings that live interviews can provide.

A researcher conducts an interview with a nurse practitioner to learn about her experience using medical device instructions.

When you conduct your user research, be sure to include representatives of all the distinct types of people who will use the product whose instructions you are developing. We consider it best practice to include about the same number of people in each distinct user group, as opposed to interviewing more people from a major distinct group and fewer people representing a minor distinct group or groups. This is to say, be inclusive, and do not develop instructions just for dominant or "average" users. Make sure you accommodate people who have special and minority needs as well.

It's also appropriate to use your professional judgment to identify needs that might not have surfaced in research but are evident based on your experience. In many cases, this approach helps identify essential requirements.

Once you learn about user needs (i.e., capture the "voice of the customer"), you can convert them into requirements for the instructions you are developing. One could write a book on this topic, and it's been done by one of this book's authors and other colleagues. This guidance is summarized in "Establish document requirements" on page 54.

Here are some key questions you might ask during user research interviews and some hypothetical replies from one interviewee.

Question	Hypothetical response (written as a user might speak)
What is your preferred means to learn to use a [medical product type]?	I like it when somebody shows me how to use a product. But it's also nice to have some simple instructions if nobody is available to do that.
In general, how do you feel about following instructions for using a [medical product type]?	I like to have them available, but I probably only read them if they look like they're not too complicated.
Do you have a preference to read instructions on paper or online?	Definitely prefer paper instructions but I worry about them being available when I need them and also being up-to-date.
What did you like most about instructions you've used for other [medical product type]?	I like when there are colorful drawings that show me what to do.
What do you like least about instructions you've used for other [medical product type]?	Too many warnings that just try to scare me. I usually ignore those.
Do you usually prefer instructions to be detailed or limited in detail?	Maybe something in between. Medium detailed, but without details that don't actually help me use the product.
Upon what aspects of using a [medical product type] should its instructions focus?	Definitely getting started the first time you try to use the product. That's when you need the most help. I like the guides you see titled *read me first*.
Is there information you would recommend excluding from the instructions?	I don't like to see all of the details that a biomedical engineer might want to read but does not help me. I need to know how to use the product and if something goes wrong, I'll call a biomed or just get another device.

Question	Hypothetical response (written as a user might speak)
Can you think of specific challenges you've experienced when reading instructions for other [medical product type]?	Yes! One time I got a strange alarm that I hadn't seen before and I couldn't find it in the instructions. I wish they included a simple table or listing of all possible alarms and the associated troubleshooting steps.
How do you feel about including illustrations?	Like I said earlier, I love drawings as long as they are clear. They should focus on what matters most.
When a whole product or specific feature is shown in an illustration, would you prefer to see a photograph or a drawing?	I think a good drawing might be best because you can make things look real but simpler, and perhaps add some emphasis or directional arrows.
Where do you usually keep instructions for [medical device type]?	I actually prefer to put the instructions up on the wall, right next to the device, so that I can always refer to it during use. That's why it's important that the instructions can be flattened easily. If the instructions are in a booklet, it's more difficult to put them on the wall.

Document your research results thoroughly and let them power the subsequent stages in your development process. This approach helps to protect against people with strong opinions and power within a development organization from disregarding the real users' needs and dictating needs that might be their own and quite possibly off-the-mark from those of the intended user population.

15

Consider the product's shortcomings

When deciding what to cover or emphasize in a product's instructions, identify tasks that might be more challenging or prone to error.

As a starting point, get your hands on the actual product or prototype and judge for yourself. Better yet, review the results of any usability tests or equivalent product evaluations to identify actual difficulties or challenges intended users encountered when using the product.

In the case of an over-the-counter blood pressure monitor, here are some tasks that might be relatively difficult:

- Wrapping the cuff around the arm properly and ensuring the Velcro surfaces overlap sufficiently to remain secure when the cuff inflates

- Quickly deflating the cuff if it causes undue pain

- Recalling past blood pressure measurements from the product's built-in memory

Knowing these tasks are challenging, you can make sure to clearly explain how to perform them properly. It's not as if you would have given these tasks only a half effort otherwise, but there is something to be said for knowing where special attention might be warranted.

A person measuring their blood pressure using a blood pressure monitor.

In the case of the blood pressure monitor, you might decide to do the following:

Present the cuff wrapping instructions as two separate illustrated steps instead of one. For example, step 1 can show the user's thumb holding the cuff against their inner arm to facilitate wrapping the cuff around the arm, while step 2 shows the user pressing the Velcro portions together to ensure the cuff stays in place when inflated.

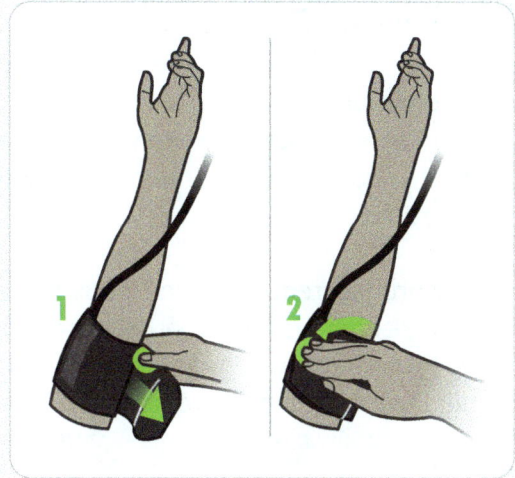

Include a visually conspicuous alert that illustrates and describes how to deflate the cuff at any time, perhaps by pressing the off button.

REMINDER

To deflate cuff, press red **STOP** button.

Along with narrative instructions for displaying previous blood pressure readings, show sample screens displaying the sequence of necessary steps.

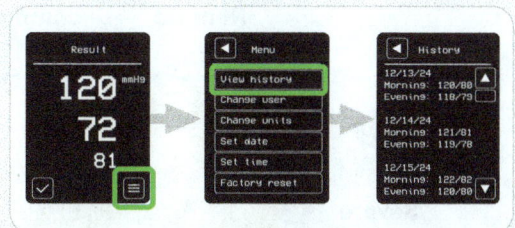

Like a helpful educator, instructions can also provide tips or describe best practices for facilitating tasks. In the case of the blood pressure monitor, the instructions might suggest wearing clothes that are loose enough to allow secure cuff placement or sitting in a comfortable position that makes it easier to hold the cuff in place when wrapping it around the arm. Such suggestions can be included in the main instructions or in adjacent sidebars (see "Sidebars" on page 131).

Consolidated troubleshooting tips can also help users recover from roadblocks that might be due to the product's known problems or many other possible causes (see "Help users resolve problems" on page 109).

16

Identify the product's use-related risks

Conducting user research and identifying the product's shortcomings will provide you with a generous list of items to address in the instructions. For most consumer products, that might be enough. But for medical products, you need to take it one step further.

Medical product regulators and industry standards call for developers to perform a use-related risk analysis for higher-risk products (e.g., Class II and III products as defined by the US FDA). The work requires identifying potential use errors and rating them in terms of their likelihood of occurrence and the severity of harm that could occur. This results in a list — usually a lengthy one — of use errors.

Each of these use errors are given a risk priority number (RPN) or equivalent. The RPNs are usually the product of multiplying the two numerical ratings that represent the risk's likelihood and severity of harm. For example, multiplying a rating of 3 out of 5 for likelihood (1 = improbable, 5 = frequent) by a rating of 4 for severity of harm (1 = marginal, 5 = catastrophic) to arrive at an RPN of 12. The RPNs can be plotted on a matrix to illustrate which risks are high and fall into a range that makes them suitable for risk mitigation.

LIKELIHOOD

SEVERITY		Remote 1	Uncommon 2	Occasional 3	Probable 4	Frequent 5
Catastrophic	5	Moderate	Moderate	High	Critical	Critical
Severe	4	Low	Moderate	Moderate	High	Critical
Moderate	3	Low	Moderate	Moderate	Moderate	High
Minor	2	Very low (Acceptable)	Low (Acceptable)	Moderate	Moderate	Moderate
Negligible	1	Very low (Acceptable)	Very low (Acceptable)	Low (Acceptable)	Low (Acceptable)	Moderate

■■ = Acceptable risk   ■■■ = Requires mitigation

Various rating scales may be used and result in a risk rating matrix such as this one, and you may use RPNs or simply describe risks on a range from low-to-critical with many or few increments.

In practice and in line with guidance from regulators, all use-related risks that could cause significant harm are subject to risk mitigation, regardless of their likelihood of occurrence. This approach protects against a particular use error being overlooked in terms of risk reduction simply because the error is judged to be highly unlikely. As such, conservatism rules when it comes to use-related risk analysis, a low RPN (based on low likelihood of use error occurrence) notwithstanding.

There are many techniques to identify potential use errors and we suggest you use multiple approaches to generate a comprehensive set. Here is a summary of some primary techniques you may choose to apply.

- **Analyze the tasks you expect people to perform using the product**. Deconstruct the tasks into discrete steps, including perceptions, mental processing steps (i.e., cognition), and actions. Then, you can identify potential errors simply through inversion. For example, if the mental processing step is to multiply a drug delivery dose rate (75 mg/hour) by a time period (2 hours) to get a total dose of 150 mg, then the use error (say determining an incorrect total dose of 125 mg) would be "User miscalculates total dose delivered."

- **Conduct a hazard analysis**. This entails identifying all potential sources of harm associated with the medical product and the use environment that could cause harm, and then working backwards to identify use errors that could lead to exposure to the harms. For example, if a possible harm is a physical burn during the use of an electrosurgical device, an associated use error might be "Touched the active electrode."

- **Conduct a known problems analysis**. This technique calls for an examination of records or past use errors that lead to adverse events and complaints. Good resources include the MAUDE* (Manufacturer and User Facility Device Experience) database and a manufacturer's own customer complaint records. From a review of such records, you simply add cited use errors to your growing list of all identified use errors.

These three analyses can also be combined with user interview and usability test findings to create a robust list of potential use errors.

Using several methods to identify use errors typically produces some overlapping results, and that's a good thing. It suggests that your overall approach to use error identification is comprehensive and starting to produce diminishing returns. It would be far worse to take an overly lean approach to use error identification and miss an area where users need more support.

---

* MAUDE - Manufacturer and User Facility Device Experience. U.S. Food & Drug Administration. Available at https://www.accessdata.fda.gov/scripts/cdrh/cfdocs/cfmaude/search.cfm

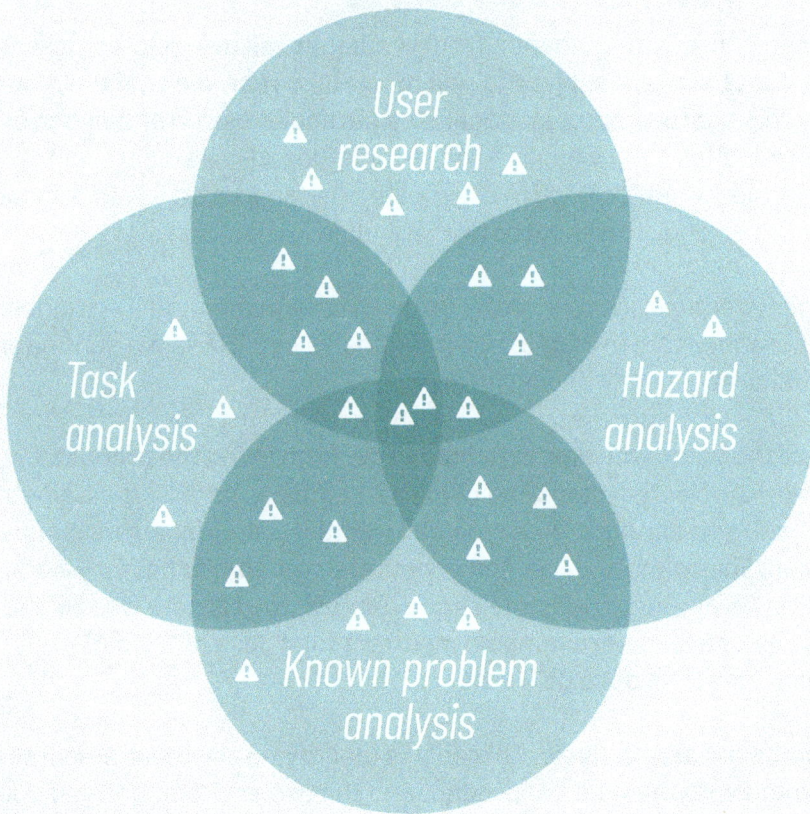

*Each use error identification method is likely to produce
some unique findings and some that overlap.*

Now, consider the case of identifying potential use errors associated with an over-the-counter glucose meter. You might be able to identify well over one hundred potential use errors.

Here are just some of the use errors that might occur when handling test strips:

- Placing too much blood on test strip
- Placing too little blood on test strip
- Placing blood on wrong part of the test strip
- Inserting test strip upside down into glucose meter
- Inserting wrong end of test strip into glucose meter
- Inserting test strip into glucose meter's USB-C port
- Using test strip from supply that passed its expiry date
- Using wrong brand of test strip
- Bending test strip while inserting it into glucose meter
- Contaminating blood sample while handling test strip

- Prematurely removing test strip from glucose meter
- Not removing used test strip from glucose meter
- Not disposing of test strip properly

So, how does a use-related risk analysis relate to designing instructions? The answer is directly.

One of the most important things instructions can do is help prevent potentially harmful use errors. Effective instructions do this by guiding people to perform tasks correctly (see "Take it step-by-step" on page 90) and by giving fair warning to users to avoid certain mistakes (see "Warn users about potential risks" on page 106).

Referencing a thorough use-related risk analysis helps you to anticipate worst-case scenarios, not just the route procedural steps, and develop instructions that can help users avoid harm.

17
Review regulatory guidance

Recognizing the critical role that instructions play in the safe and effective use of medical products, regulators have complemented their requirements with guidance for instructional content. Reviewing this guidance is a key step in the development of medical product instructions.

In "Instructions must meet regulatory requirements" on page 13, we summarize the FDA's regulatory requirements for medical device instructions. In recent years, the FDA has complemented their regulatory requirements with guidance documents that are not legally binding but give a good sense of the FDA's expectations for instructional documents.

The FDA's guidance for instructions varies by product type. Below, we provide a brief description of each product type and an overview of the FDA's expectations.

Guidance for combination drug products
Combination products, such as auto-injectors and pre-filled syringes, which include both a device and a biologic or drug, are reviewed by the FDA's Center for Drug Evaluation and Research (CDER).

In July of 2022, CDER published *Instructions for Use — Patient Labeling for Human Prescription Drug and Biological Products — Content and Format: Guidance for Industry.** As the title implies, the document provides detailed guidance for both the content and formatting of combination product instructions for patients, with the goal of increasing consistency among all combination product instructions. In turn, this high level of consistency is intended to provide patients with a consistent presentation of information that facilitates correct use.

Much of CDER's general guidance is consistent with recommendations from this book, such as:

- Use simple, non-technical language and write at or below the national average reading level (see "Use plain language" on page 139)

- Use active voice and imperative sentences beginning with a verb (see "Use active voice construction" on page 150 and "Permissive versus prohibitive phrasing" on page 152)

- Avoid abbreviations that might be misinterpreted by users

---

* Available on FDA's website at https://www.fda.gov/media/128446/download

- Utilize and standardize headings to improve readability and navigation (see "Make instructions visually appealing" on page 117)

- Use a sans-serif typeface and a font size of at least 8 points (see "Sizing and styling text" on page 124)

- Present instructions step-by-step (see "Take it step-by-step" on page 90)

- Use clear visuals (see "Make illustrations clear" on page 160)

- Employ sufficient white space (see "Make instructions visually appealing" on page 117)

Some of CDER's formatting guidance is more specific and prescriptive than some of the recommendations we provide in this book, such as:

- Use uppercase letters to highlight specific information, specifically the title of the document (e.g., INSTRUCTIONS FOR USE) and the drug name (e.g., PAXLOVID), while limiting the use of all-uppercase words elsewhere

- Use title case for all headings

- Ensure the presentation of the drug name (except in the Product Title) matches its presentation in the prescribing information (PI) typically presented before the instructions

- Limit bolding to the title, product title, pronunciation spelling, dosage form, route of administration, the controlled substance symbol, headings, step numbers, and figure titles, with occasional additional bolding and underlining to highlight key information

- Use only black type on a white background to maximize legibility and ensure all colored text and visuals translate well to black-and-white or grayscale printing

CDER is similarly prescriptive regarding the contents to be included in the instructions and suggest that instructions include the following contents in the presented order:

- **Title:** INSTRUCTIONS FOR USE in bold, upper-case letters centered at the top of the page

- **Product Title:** The product's proprietary name in upper-case letters followed by the pronunciation spelling in brackets, nonproprietary name in lowercase letters and in parentheses, dosage form, route of administration, and (if applicable) the controlled substance schedule

- **Purpose Statement:** Text that states, "This Instructions for Use contains information on how to [insert application verb] [insert drug name]"

- **Visual of Drug Product:** An image of the product and, if applicable, all constituent parts (e.g., vials and syringes that come in a reconstitution kit)

- **Important Information for Patients:** Information about specific actions patients must take to prepare, administer, store, and/or dispose of the drug, presented under a heading that states "Important Information You Need to Know Before [Application Verb] [Drug]"

- **Preparation Instructions:** Under a heading that states "Preparing to [application verb] [drug]," include all steps required for the user to take prior to administering the drug

- **Administration Instructions:** How to administer the drug under a heading that states [application verb] [drug]

- **Disposal Instructions:** Describes any specific disposal instructions for the product under the heading "Disposing of [Drug]"

- **Additional Information:** At the bottom of the last page, include 1) resources or additional information, 2) the name and address of the manufacturer, packer, and/or distributor or the name, address, and license of the manufacturer and, if applicable, distributor, 3) a statement that says, "This Instructions for Use has been approved by the US Food and Drug Administration" with either "Approved: Month YYYY" or "Revised: Month YYYY"

Guidance for medical devices

Medical devices, which are intended to diagnose or treat medical conditions, are subject to requirements set by the FDA's Center for Devices and Radiological Health (CDRH). In 2001, CDRH published the document *Guidance on Medical Device Patient Labeling; Final Guidance for Industry and FDA Reviewers.*[*]

Given the diversity of medical devices, which can range from relatively simple products like thermometers to complex devices like MRI machines, CDRH's guidance for instructions is less prescriptive than those provided by CDER for combination product instructions.

---

[*] Available at https://www.fda.gov/media/71030/download

However, many of the recommendations are helpful and consistent with those we provide in this book, such as:

- Include a table of contents and glossary in lengthier instructions

- Write at or below an eighth grade reading level

- Use plain language, short sentences, and the active voice

- Define technical terms when you first use them in the instructions

- Present the most important information near the start of the instructions

- Group similar information together

- Use simple graphics and illustrations to demonstrate concepts and interactions with the device

- Place graphics next to the associated text

- Use highlighting, bolding, and other visual design techniques to emphasize key points

- Make warnings prominent and noticeable

- Create warnings that include signal words, hazard statements, consequences, and instructions for avoiding the hazard

- Test the instructions with target users to ensure comprehension and usability

Note that although both CDER's and CDRH's guidance is focused on instructions intended for patients, many of the suggestions they include are equally applicable to instructions for healthcare professionals. If you are designing instructions intended only for healthcare professionals, you will still be well served to use the FDA's guidance as a starting point.

Also, be aware that FDA occasionally updates its guidance and publishes new guidance documents. Be sure to check CDER's or CDRH's websites for the latest guidance when you begin your instruction design effort.

18
Establish document requirements

Developing instructions is certainly a creative exercise. However, it helps to start by establishing some document requirements. Such requirements can provide structure for the creative work to follow and are likely to produce more effective instructions. The same can be said for creating a good hardware or software user interface.

Requirements can help instructional designers in myriad ways, including these major ones:

- Identify the required information at a general level
- Identify opportunities for reducing use-related risk through instructions
- Identify any document production constraints
- Specify the target audience and their minimum reading level
- Describe characteristics that enhance legibility and readability
- Describe features that will help users navigate the document

Generating useful requirements typically involves the following steps:

1. Conduct user research to better understand the prospective users and their needs for the product
2. Identify applicable regulatory requirements for the instructions' content and formatting
3. Review instructions from predecessor and comparable products
4. Review best practices for writing requirements
5. Develop an outline for the requirements
6. Write draft requirements
7. Seek feedback on draft requirements
8. Finalize the requirements
9. Adapt the requirements as needed

Developing requirements is an iterative process. You'll likely revise your draft requirements and seek feedback from colleagues as well as prospective users multiple times before you can finalize them.

In later product development stages, you might also need to revisit these requirements. For example, you might discover during usability testing that the instructions do not address a potential use error that wasn't previously considered. Similarly, evolving production or budget-related constraints might require you to deviate from your originally preferred format or paper type.

In such cases, changes to the document requirements may be in order. By revising the document requirements rather than discarding them, you ensure that the rationale for the changes are kept as part of the design history and that all members of the development team (as well as key stakeholders) understand the changes.

Best practices

Having document requirements is an essential step towards ensuring the quality of instructions. But, producing high-quality instructions begins with having high-quality requirements. In particular, effective requirements tend to be:

- **Clear**: Easy to understand with minimal possibility of misinterpretation due to ambiguity in the requirements

- **Complete**: Cover all key aspects of the instructions, including its content, formatting, language, visual design, and production constraints

- **Consistent**: Do not conflict with other requirements that you have established for the instructions

- **Document-centric**: Define characteristics of the document, not the people who will be using it

- **Verifiable**: There should be a way for you to inspect the instructions and objectively confirm if the requirement is met

- **Traceable**: Where appropriate, the requirements may need to refer to related product objectives, such as marketing requirements, production needs, or use-related risks

The following table provides specific examples of ineffective document requirements, along with explanations of their shortcomings and examples of improved requirements that are more effective.

Original requirement	Why is it ineffective?	Improved requirement
The instructions shall be easy to follow.	"Easy to follow" is subjective and open to interpretation, which makes it difficult to address and verify.	The instructions shall feature numbered headings that describe each section's contents with a maximum of five words.

Original requirement	Why is it ineffective?	Improved requirement
Users shall have a high school education to read the document.	The requirement sets a criteria for the user, rather than for the document.	The instructions shall have text appropriate for a Grade 9 reading level, as measured by the Flesch-Kincaid reading score.
The instructions shall be printed on the brightest and highest quality paper possible.	The requirement is open ended and could result in the use of materials that are more expensive but not well suited for use (e.g., glossy paper that is difficult to read in certain lighting conditions).	The instructions shall be printed on materials that have a minimum brightness rating of 90, as measured by the ISO 2470-2 standard for paper brightness.
The instructions must be exhaustive and concise.	The requirement sets contradictory goals that cannot be met simultaneously.	The instructions shall cover all key use scenarios, including setup, daily operation, maintenance, and troubleshooting. The instructions shall describe each use scenario concisely, using a maximum of 10 steps and/or 500 words.
The instructions shall be legible in all lighting conditions.	The requirement sets an unrealistic demand that cannot be met.	The contrast of text to background shall be at least 7:1 as measured in an environment with ambient lighting of 50 lux or higher.

19

Determine production opportunities and constraints

An instructional designer's imagination may be unlimited, but all instructions are constrained by their production factors: the processes and characteristics that form the instructional "product." Having a good understanding of the production requirements helps avoid usability shortcomings and inspire creative design solutions.

When sizing and siting a home, an architect must consider building site constraints. There will be setbacks, such as zoning requirements, driveway access considerations, trees to preserve, and perhaps rocky outcrops to avoid. In addition, an architect must consider how the materials available for construction will affect their design. Factors such as the materials' structural strength, weight, thermal properties, fire resistance, and cost can all affect an architect's design decisions.

And that's before they even think about the views from the living room!

What are the "site conditions" and "material factors" you need to consider when designing instructions? Here is a mix of self-evident and less obvious considerations, many of which we explore in more detail in the Design section of this book:

Architects review landscape and environmental factors to inform their building design.

Printed instructions

Cost. Determine the target cost of goods for the instructions and let this guide subsequent decisions. Of course, you can advocate for the funds necessary to produce a good product with budgetary authority.

Offset vs. digital printing. Offset printing, also known as offset lithography, is a more traditional printing method that prints using custom-made metal plates, while digital printing uses similar technology to that found in common home or office printers. Offset printing produces higher quality prints but is usually only cost effective when printing thousands of copies.

Print run. Estimate the number of copies you'll print, which can have a direct effect on the cost per copy. As you might expect, printing more copies can significantly reduce the price per copy, especially when using offset printing technology.

Paper size. Utilizing industry-standard paper sizes tends to be more cost effective than using custom-cut sizes. Estimate the amount of content you need to fit on a page/sheet to determine whether standard sizes will be a good fit (see "Select orientation (aspect ratio)" on page 84).

Paper thickness. Determine the paper thickness and account for bleed through if it is unavoidable due to cost constraints (see "Paper thickness" on page 180).

Folding. Determine the optimal folding patterns, which you can take into consideration when laying out content (see "Fold lines" on page 182).

Bleed edges. Determine if the document can be printed with bleed edges, which allow content such as a background fill to cover the edges of the page rather than having a white margin at the edges of the page.

Full bleed　　　　　**No bleed**

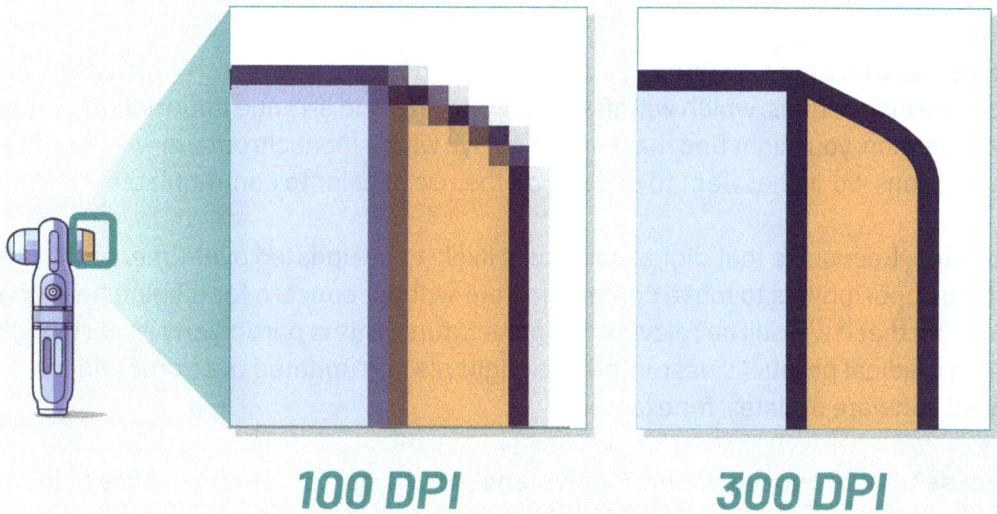

100 DPI   **300 DPI**

Resolution. Determine the target resolution for content, which is ideally ≥ 300 dots per inch (DPI) so that there is no evident jaggedness or blurriness.

Color vs. monochrome. Determine if the instructions can be effective in monochrome (i.e., black and white) or if color printing is required. If color printing is an option, explore the tradeoffs of using full color printing (typically using the four colors: cyan, magenta, yellow, and black) or limited color printing (using a combination of two or three colors in addition to black).

Paper coating. Determine if the document will be coated or even laminated to prevent contamination in case it influences how you (a) size the document in view of the use environment and (b) how you fold it (or do not fold it).

Graphics. Consider the format of any graphics in your document. Vector graphics scale smoothly to any size. However, images that are comprised of pixels (i.e., raster images) will become blurry when enlarged. The solution is to assure that the graphic will be at the aforementioned ≥ 300 DPI resolution when printed.

Photographs. When including photographs in your documents, use uncompressed formats such as Tag Image File Format (.TIFF) or Photoshop Document (.PSD) to ensure the images remain sharp no matter how many times they are saved. Compressed formats, such as JPEG and PNG, can lose image quality every time they are re-saved.

Digital instructions

Screen size and type. Determine size, format, and type of screens on which users might view the instructions, which will affect how you size and arrange information. Although less common, you might find that some displays will be monochrome, in which case your instructions would not want to depend on the use of color to communicate

Updates. Recognize that digital content is likely to be updated over time, so this might create opportunities to make content current without concern for it being "evergreen," meaning that it will still be relevant far in the future. This is particularly important given that a medical product's user interface might also be updated over time, via internet-based software updates, for example.

Accessibility. Acquaint yourself with standards for digital content accessibility, such as those published by the World Wide Web Consortium.*

Animations. Desktop and laptop displays can readily display animations, as can a display embedded in a medical product, such as a dialysis machine. Therefore, you may consider if some instructions should be complemented or supplanted by animations.

Voice. Some medical device users like voice instructions. As such, you might wish to explore ways to complement the instructions with audible prompts or guidance.

Baxter's Amia automated peritoneal dialysis system provides onscreen instructions in the form of text, graphics, animations, and voice. (Image provided courtesy of Vantive Health LLC)

---

*   W3C Accessibility Standards Overview. World Wide Web Consortium. https://www.w3.org/WAI/standards-guidelines/

20
Develop preliminary instructions

Product development teams tend to begin their design and development efforts with preliminary activities such as wireframing and prototyping, which enable them to iteratively evolve toward a finished product design. Instructional designers are well-served to take a similar approach to their work and develop preliminary instructions as early as possible during the product development effort.

In "Good instructions are developed as an integral part of the product" on page 9 we highlight some of the benefits of developing preliminary instructions early in the product development effort. In this section, we'll describe the process of creating preliminary instructions.

Calling instructions preliminary is really another way of saying that they're incomplete. Preliminary instructions can be incomplete in one or more of the following ways:

- Including only a subset of the content that will be in the final instructions

- Containing meaningless placeholder text, also known as "lorem ipsum" text, to approximate the amount of written content that might be included in the document

- Containing unrefined drafts of instructional text or even just high-level descriptions of text that will eventually be included (e.g., "Instructions for placing protective cover over needle")

- Having empty blocks in place of graphics that will be added in the future

- Using informal photos to depict interactions with an early prototype of the product

- Using rough illustrations in place of more refined illustrations that will be prepared later in the development process

- Utilizing a format that is simplified or altogether different from the format that will be used in the final instructions

- Including blank placeholders for sections that will be added in the future

You can create preliminary instructions by hand, sketching out rough designs and writing placeholder text on different sizes of paper to get a sense for their feel. Shortly thereafter, you can create early digital mockups, similar to what user interface designers call wireframes, which contain a visually unrefined arrangement of placeholder content.

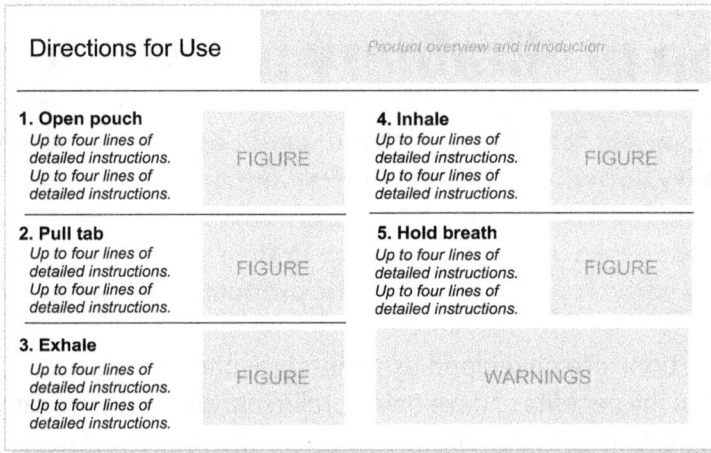

Directions for Use	Product overview and introduction

1. Open pouch	4. Inhale
Up to four lines of detailed instructions. Up to four lines of detailed instructions.	FIGURE	Up to four lines of detailed instructions. Up to four lines of detailed instructions.	FIGURE

2. Pull tab	5. Hold breath
Up to four lines of detailed instructions. Up to four lines of detailed instructions.	FIGURE	Up to four lines of detailed instructions. Up to four lines of detailed instructions.	FIGURE

3. Exhale	
Up to four lines of detailed instructions. Up to four lines of detailed instructions.	FIGURE	WARNINGS

An example of preliminary instructions that convey the intended information arrangement, but lack detailed content.

By definition, the preliminary instructions will evolve and change over the course of product development. But, having the preliminary instructions in hand will make it easier for you to make certain design decisions later in the instruction development process.

For one, preliminary instructions give you some sense of the amount of information that will be required in the final document. At a minimum, it gives you at a starting estimate on which you can begin to make future design and production decisions. For example, if your preliminary design exercises reveal that the instructions will depict only a handful of actions that users can perform, you can probably assume that your final instructions will be a short document rather than a lengthy manual. Conversely, if you're struggling to limit the amount of content in your preliminary instructions, you might discover the need to produce a more extensive manual.

Making these determinations early in the instruction development effort brings two more benefits. One is a logistical benefit: you'll have more time to identify and work around any production constraints resulting from your preliminary design exploration (see "Determine production opportunities and constraints" on page 57). The second is a creative one: creating "quick and dirty" preliminary designs enables you to explore alternative design directions without getting too attached to any particular solution. This exploration of divergent options is more likely to lead to a final instructional design that is well suited to users' needs (see "Explore multiple concepts" on page 63).

21

Explore multiple concepts

Most experienced designers will agree that the best way to arrive at an effective design is to generate many divergent concepts and then converge on a preferred one.

In the industrial design world, some people call it the sandwich: you develop conservative and radical solutions, as well as one that splits the difference. By looking at concepts on the edges, you learn a lot about what might be the best design solution. Often, the "compromise solution" is a hybrid of the middle solution and some features from the edgy solutions.

Alternative design concepts may also be variants on a basic solution, whereby a particular solution showcases a particular feature or style but starts with the same foundation. Automobile designers often do this on the pathway toward producing a leading concept car.

Medical device instructions are no different. You're more likely to produce an optimized final design if you start the design process by exploring divergent initial concepts.

The original Ford Mustang evolved substantially from early prototypes (top) to its final design (bottom).

The alternative design concepts you create can vary in multiple ways, including the instructions' organization, writing style, formatting, layout, and illustration style. We suggest incorporating bolder, as opposed to minor or subtle, differences among your concepts. The illustrations below show a variety of design concepts for the same instructions on how to monitor your heart activity using a portable ECG monitor intended for home use. The illustration on the next page shows a potential hybrid solution that could emerge as a preferred concept.

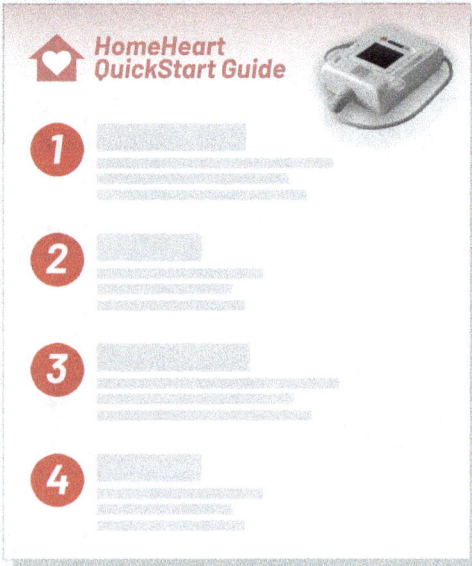

When it comes time to choose a preferred design direction, be mindful of how you mix and match design elements from the original design concepts. Much like creating a harmonious meal, blending design ideas requires a thoughtful approach to preserve the essence of each ingredient. For example, one concept's streamlined organization might not be as effective if it's combined with a more technical writing style from an alternative design concept. By cautiously evaluating the compatibility of various concepts and thoughtfully blending their features, you can create a cohesive, effective solution that maintains the essence and strengths of the original ideas.

22

Conduct usability tests

Usability testing is a cornerstone of human factors engineering, a discipline with activities required by many international regulators responsible for the evaluation and safety assurance of medical devices and combination products.

A usability test brings products and representative users together in the context of a use scenario, a task or series of tasks (i.e., steps) that users would be expected to perform in real-world conditions to use the device. As the participants perform various tasks comprising the overall use scenario, test personnel look for indications of difficulty and outright mistakes (also called use errors). Through this process, device manufacturers are able to identify potential pain points and opportunities for improvement in their design (from what is known as a formative usability test) or to demonstrate the safety and effectiveness of their final design (from what is known as a human factors validation test or summative usability test). Devices frequently undergo multiple formative usability tests with the goal of each design moving closer towards the final, safest design concept that is most appealing to use.

A usability test participant reads instructions for using a glucose meter while a usability engineer observes.

As we have stated elsewhere in this book, the instructions accompanying the product can have a positive (or negative) effect on user performance, and the overall quality of instructions can impact a usability test. Effective instructions help the user progress smoothly through tasks. Conversely, poor instructions hinder user performance when the evaluated product is unintuitive to use or when complicated tasks arise. Additionally, instructions can be the main focus of usability tests, where the test personnel focus on the user's ability to follow and understand the instructions rather than to use a device, to assess the overall effectiveness, clarity, and design of the instructions.

You might wonder if it's sufficient to just inspect instructions for design flaws rather than investing resources in conducting a usability test. The simple answer is that an inspection might suffice, if the inspector is very good at detecting design flaws. But, you'd be taking a chance of missing something important.

We'll enlist the watering can shown below, an example of the architect Katerina Kamprani's "deliberately inconvenient everyday objects," to reinforce this point.

The Uncomfortable Watering Can
(© The Uncomfortable by Katerina Kamprani)

When inspecting this watering can for design flaws, the spout is an evident problem from a functional perspective. However, what is less obvious from simple observation, and would become evident through usability testing, is that the handle's circular cross section and distance away from the container make it uncomfortable to hold and manipulate.

But why couldn't an inspector fill a prototype with water and use it? They certainly can and do. In this case, the inspector becomes a representative user, or test participant, in an informal usability test. But let's say the inspector has particularly large hands and high level of strength, making the design of the handle a non-issue. However, another person with smaller hands and less strength falters when using the watering can. A usability test would yield additional data points (i.e., feedback from test participants) to justify changing (or not changing) the handle's design to one that meets the needs of the intended users.

We do not want to minimize the role of design inspection when developing devices (or instructions). Frequently, the two complement each other. In the example of the watering can above, a design inspection would allow the manufacturer to rectify the spout's flaws prior to conducting usability testing, enabling representative users to formulate feedback on the device's design that is specific to their experience using the device and not universal design flaws.

While most people might be well qualified to judge a common watering can, this is not the case with many medical devices. For example, someone developing instructions for an insulin pen-injector would likely be intimately familiar with how the device works and might not have any dexterity or visual impairments. Contrast this description to that of an actual user, a person with type 1 diabetes who has injected insulin using a syringe and vial for 25 years, might have some retinopathy affecting their vision or neuropathy affecting their dexterity, and is using a pen-injector for the first time. This kind of person will bring a different perspective to the task that affects how they will perceive, use, and judge the experience of delivering insulin with a pen-injector.

At a high level, the usability testing process usually includes the following steps:

1. **Determine the test purpose.** A usability test's purpose might be to solicit user feedback on alternative designs, assess user performance on a design iteration to ensure safety and efficacy, or even collect data for submission to regulatory bodies, among others. Determining the desired outcomes, goals, and scope of the usability test at the outset is imperative and informs later steps.

2. **Develop a test protocol.** A test protocol describes all relevant aspects of the test, including the objectives, participants, environment (real or simulated), use scenarios, and performance measures (e.g., correct completion, time on task, number of use errors, subjective attribute ratings). Test protocols typically call upon test participants to use a medical device and, if they choose, the accompanying instructions. However, a usability test might need to include knowledge tasks that call upon test participants to read and interpret safety-critical content in the instructions. Coming to agreement on the protocol with the primary stakeholders in the medical device and instruction development effort helps ensure buy-in regarding the test results and any design updates that might be necessary based on test findings.

3. **Conduct a pilot test and revise the protocol as necessary.** Conducting a pilot test serves to get the kinks — or artifacts — out of the testing protocol before you start collecting data that gets reported to the stakeholders and possibly even the medical device regulators. After a pilot, you can revise the test protocol as needed to remove these artifacts.

4. **Recruit participants.** When preparing a submission to the US FDA, the final test (i.e., human factors validation test) must include at least 15 representatives of each distinct user group. As such, a final test might need to include 15 laypersons/patients and 15 registered nurses, if these are the types of people who will interact with a given medical device. Many other countries have a comparable standard for validation tests. In contrast, when conducting a formative usability test, you can

include the number of test participants that you consider most productive. Often, a sample of 5-12 people will suffice.

5. **Establish the test environment.** Testing should occur in an environment that has approximately the same performance-influencing characteristics (e.g., lighting, noise, distractions) as the product's real-world use environment. For example, a device that might be used in an ambulance likely requires a small working area with traffic sounds and an unsteady work surface. In contrast, a device intended for use in an operating room would require bright, overhead lighting in the test environment and perhaps other people acting as the representative user's colleagues collaborating on a patient's procedure. As a third example, a device intended for home use might require dimmer lighting and television noises to simulate the use environment.

6. **Conduct the test sessions.** Testing engages the participants in selected use scenarios, which may include many steps or tasks. Use scenarios should include steps or tasks that you have determined to be safety-critical as well as those central to the device functioning properly and satisfying users. Using a blood pressure monitor as an example, the first use scenario could be preparing to use the monitor for the first time, starting with taking it out of the box. While not stated explicitly to the participant, the scenario might require entering one's name, body weight, and the date into the device's memory. The second use scenario could be taking one's blood pressure. The third use scenario could be recalling a blood pressure measurement taken weeks ago from the device's memory. Any one of these steps, if performed incorrectly or not at all, could prohibit a user from taking their blood pressure or assessing trends in their blood pressure, preventing them from seeking further care if needed.

7. **Consolidate and analyze the data.** Test data may include the time it took to complete a task, the degree of use scenario completion, the number and type of use errors, difficulties the user experienced when completing the task, close calls where the user almost made a mistake then self-corrected, attribute ratings, comparative rankings, and participant comments (what some people call anecdotal remarks). Your post-test data analyses will depend on the data you collect and might include analyzing quantitative data (noting that typical usability test sample sizes do not yield statistically significant results) or identifying patterns in the test participants' comments. Perhaps the most pertinent analysis, known as the root cause analysis, is to hypothesize the root causes of observed interaction problems and, most importantly, the root causes of safety-related use errors (i.e., critical mistakes).

8. **Report the results.** The results of all usability tests should be well documented. The reports become an integral part of a medical device's design history file. They also enable you to describe to regulators the journey from a preliminary design solution to a final one that will enable safe and effective medical device use, supported by well-designed, validated instructions. It might be sufficient to document a formative usability test of the instructions in a memorandum or slide presentation, but virtually all human factors validation test reports take the classic form of a narrative report augmented by figures, tables, and appendices.

As alluded to throughout this section, there are two primary forms of usability testing used during a device's development: formative usability testing and human factors validation testing. At a high level, formative testing is completed on devices still in development to identify the current design's strengths and relative weaknesses and might be conducted on several iterations of the design. The human factors validation test is typically conducted once on the production-equivalent design.

Here are some quick tips about conducting formative usability tests and human factors validation tests:

- You might want to conduct a preliminary (i.e., formative) usability test that requires test participants to follow the instructions, thereby ensuring that the instructions are assessed comprehensively. You'd be doing this with the understanding that in real-world use, people might ignore the instructions or use them sparingly. As such, requiring the use of instructions is typically not included as part of the human factors validation test.

- While formative usability testing is not explicitly required by medical device regulators, there is an implicit need for it because regulators are eager to see human factors engineering applied early and throughout the development process and want manufacturers to report the results of their formative evaluation work. Human factors validation testing is explicitly required by regulatory bodies, and there are several guidance documents that should be followed when designing these tests.

- It's common for medical device developers to conduct 2-3 formative usability tests ahead of human factors validation testing. You want to approach validation testing with high confidence that things will go well based on the results of the prior formative usability tests, which are likely to be smaller scale tests.

To learn more about usability testing, see "Additional resources" on page 204.

23
Finalize the instructions

Leonardo da Vinci supposedly said, "Art is never finished, only abandoned." And yet, all instructional documents must be finished, or else the users who need them will never have them. So, how can we know when instructions are finished?

In our professional practice of designing instructions, we consider instructions finished when we believe we can't make them significantly better. Specifically, we ask ourselves the following questions:

1. Can we better address opportunities for improvement identified in usability tests or other formative evaluations?

2. Can we further simplify the instructions without reducing their effectiveness?

3. Can we implement additional enhancements to make the document more accessible to individuals with impairments?

Once our answers to all three questions are negative, we consider the design of the document complete. But, finalizing the instructions also involves a final round (or rounds) of quality checks to ensure that the instructions are free of errors.

Different organizations might have standardized procedures for performing quality checks, and you should make sure that your process adheres to your organization's standards. That said, a thorough check of an instructional document's quality should at least cover the following aspects:

Technical accuracy. Given that the design of a product can continue changing up to its finalization, it is possible for discrepancies to arise between product characteristics reflected in the instructions and those existing in the actual product.

Content completeness. Certain content, such as document version numbers, manufacturer address details, and customer support numbers, are sometimes represented as placeholders until the final stages of document design. Final rounds of quality control should ensure that these details are present and accurate.

Language consistency. As a product design evolves, so do the terms that are used to refer to its components and features. Final quality checks should ensure that the product and its accessories are referred to using consistent terminology throughout the document.

Visual consistency. We discuss the importance of consistent visual styles and graphics in "Be consistent" on page 94. But more subtle visual inconsistencies, such as differences in alignment, spacing, and use of color, can also accumulate over time and should be checked for during the final quality reviews.

Stylistic consistency. As a document undergoes multiple revisions and edits, its language and visual design can evolve to reflect various contributors' stylistic preferences. Left unchecked, this evolution can result in inconsistencies that give a document an unprofessional quality.

Compliance and legal considerations. As we discuss in "Instructions must meet regulatory requirements" on page 13, different countries have different regulatory requirements for the content and format of instructions. Final quality checks provide one last chance to ensure that the instructions comply with their local requirements of the countries in which they will be used.

Note that such quality checks might need to be repeated for any document variants you produce. For example, if you create different versions of the instructions for different markets or different modes of distribution (e.g., print vs. digital), you will need to ensure that each variant meets its quality requirements before it can be considered final.

24

Verify and validate the instructions

Verification and validation (V&V) are important steps in the medical device development process that ensure a product has been *designed right* and that the developer has *designed the right thing*, which is how the FDA's human factors officials described V&V in the early 2000s.

V&V calls on manufacturers to determine if 1) they have met their own design requirements and 2) the finished product serves its purpose well (i.e., that it works in the hands of the intended users).

Nominally, a product's instructions are considered part of a product's user interface and, more specifically, its *labeling*. That means instructions are subject to V&V.

You can read more about V&V, which is part of what the FDA calls Design Controls, at the agency's website.* While there's a lot to V&V and a full explanation of the process is beyond the scope of this book, here are some key points pertaining to instructions:

- Instructions should be based in part on identified user needs, which may be derived in part from user research (see "Determine user needs" on page 40) and instantiated as design requirements (see "Establish document requirements" on page 54).

- Finished instructions should be inspected in a structured manner to ensure that all of the documented requirements related to the instructions have been met. If they have been met, you can consider the instructions to be verified. If some requirements have not been met, the options are to revise the instructions to meet them or go through a formal process of changing the requirements for justifiable reasons.

- Finished instructions should be given a road test, so to speak. That is, their usefulness should be evaluated by means of human factors validation testing (see "Conduct usability tests" on page 66). Normally, you make the instructions available to a test participant in the same manner as they would be available in a real-world scenario. Such testing reveals if the instructions cause use errors — particularly those that could lead to significant harm — or lead to close calls or difficulties. If the instructions cause interaction problems, they might need to be revised and retested.

---

* FDA Design Controls for Medical Device Manufacturers, available at https://www.fda.gov/media/116573/download

Often, as a part of human factors validation testing, a test participant might use a product without referring to the instructions or perhaps only refer to the instructions sporadically rather than systematically. In such cases, and for the sake of ensuring a thorough validation, instructions should undergo "knowledge testing" as a complement to the hands-on-product portion of the HF validation test.

During a knowledge testing portion of a validation usability test, participants read and interpret content from the instructions. Researchers then ask follow-up questions to determine if the participants' interpretations are substantially correct or not. If not, the instructions might require revision and retesting, as stated previously. As you can imagine, a modicum of professional judgment comes into play when deciding if the instructions are good as-is or not.

25

Prepare the instructions for production

One of the most exciting steps of preparing an instructional document is sending it off for production, where your digital draft transforms into thousands (or even millions) of printed copies or is shared online for global access. But, this step can also be daunting, especially for first-time instruction creators.

At first glance, preparing instructions for production seems like a straightforward step. In the case of printed documents, it involves the delivery of a set of files (either the source design files or compiled PDF files) to the printer. In the case of digital instructions, it involves the delivery of one or more files (typically HTML files) to people responsible for managing content on a company's website (typically web content managers or web administrators).

But, like a baker about to place their dough into the oven, sending instructions off for production is a point of no return. Once a document's production has begun, it's too late to change the document unless you're willing to dispose of any copies already produced. Similarly, there are many ways in which an incorrectly prepared file can lead to printing errors that necessitate costly repeats of the production process.

Fortunately, following these steps and best practices can reduce the chances of production errors or problems:

Coordinate with your production partners. Printers can have different preferences for the type and format of files that they receive for production. Make sure to ask for their preferences and requirements before finalizing documents for production. And don't hesitate to ask for clarifications regarding any terms that are unclear to you.

Provide the source design file or a high-quality compiled PDF. Instructional documents typically consist of multiple files. The primary file is often referred to as the source design file and usually includes all the content and layout details for your document. Common formats for such files include Adobe InDesign and Adobe Illustrator. Providing the source design file enables printers to verify that the document meets their technical requirements for production but introduces the risk of unintended changes before production. As an alternative, you can prepare a compiled PDF (also known as Adobe Acrobat document) that cannot be easily edited. However, if you provide a PDF to your printer, you must make sure that it meets certain technical requirements that we describe in more detail below.

Include all assets required for the document. In addition to the source design file, instructional documents typically require additional assets, or files, that contain content and resources needed to print the document correctly. Commonly used assets include the font files for all typefaces, graphic files for any illustrations or photos, and word processing files with the document's text. Note that if you compile a PDF of your document, you must embed any external assets in the PDF to ensure that they are available for printing.

Ensure that all references are correct and up to date. If your document includes references to specific pages (e.g., in a table of contents or inline references within the text), make sure that they refer to the correct sections or page numbers in your document. Many publication design applications, such as Adobe InDesign, include features to detect and correct erroneous references within a document.

Confirm and calibrate the color model. A common error when preparing documents for production is to include graphics that utilize a color model intended for display on a screen (i.e., an RGB color model), rather than one intended for print (typically CMYK or Pantone color models). Using a screen-based color model can cause your printed document to have colors that look significantly different from those you see in your digital previews of the document, resulting in inaccurate graphics or illegible color contrast combinations. Print-oriented color models enable you to define specific combinations of inks (such as cyan, magenta, yellow, and black in the CMYK color model) or standardized shades of particular hues (also known as spot colors, as defined by the Pantone color standard, for example). Using these models in conjunction with color-calibrated computer displays helps you ensure that colors in the printed documents closely match those you see in your design software.

Include appropriate printer marks. If you choose to provide your printer with a compiled PDF, you will need to include special marks in your PDF that convey technical details to the printer. Such marks, which can be generated automatically by most publication design software, may include:

- *Crop marks*. If your document uses a non-standard paper size, crop marks indicate where the paper should be trimmed to its intended size.

- *Bleed marks*. If your document features printing up to the edge of the page, your printer marks should indicate how far beyond the crop marks the "bleed area" (i.e., printed area) extends.

- *Registration marks*. Traditional offset printing of multi-color documents requires printers to use a separate color plate for each of the ink colors used to print the document. Including registration marks in your PDF enables the printer to ensure that each of the color plates are correctly aligned and minimizes the risk of "color fringing" that occurs when the plates are misaligned.

- *Color bars*. To ensure accurate and consistent colors across all the printed documents, printer marks can also include color bars that identify the base colors (e.g., cyan, magenta, yellow, and black) and any spot colors used in the document.

- *Fold marks*. If your document will be folded, the locations at which the page will be folded can be indicated as fold marks, which are typically included in specially defined layers in the PDF. These layers appear in the PDF when it's viewed in a PDF viewing application but are not actually printed.

Check for hidden content. Publication design applications allow designers to place content on multiple layers in a document, as well as hide or show certain layers as they work on the document. One consequence of this feature is that layers might remain hidden when viewed in the design application, but be erroneously included in the printed document. Conversely, a designer might inadvertently leave a layer hidden when compiling a PDF and cause content to be missing from the generated document. Similarly, in design programs, text boxes can contain overflowed text that is included but not visible.

The good news is that contemporary design software, such as Adobe InDesign, can automate many of these steps. Check your design software for features such as preflight or pre-press checks, which can help facilitate and accelerate the steps we've described.

CHAPTER THREE
DESIGN CONSIDERATIONS

26

Develop a good conceptual model

You might have heard about user interface (UI) designers developing *conceptual models* of their product.

Typically, this means:

- Thinking about the user goals and product functions that a user interface should serve

- Identifying users' general way of thinking about the product's operations (i.e., the user's mental model)

- Deciding how to organize user interface elements in way that matches users' expectations

User interface designers typically start with affinity diagramming[*] and bubble diagramming[†] exercises to identify alternative conceptual models. These techniques help organize user interface functions and sometimes general content categories into

Designers conduct an affinity diagramming exercise to identify potential content for an instructional sheet.

* R.F. Dam, and T. Y. Siang, Teo Yu. Affinity Diagrams: How to Cluster Your Ideas and Reveal Insights. Interaction Design Foundation. https://www.interaction-design.org/literature/article/affinity-diagrams-learn-how-to-cluster-and-bundle-ideas-and-facts
† Bubble Diagrams. ConceptDraw. C.S. Odessa Corporation. https://www.conceptdraw.com/solution-park/diagram-bubble

related groups. They can also reveal interrelationships among the groups, thereby suggesting navigation paths (e.g., workflows and menus).

For the instruction designer, an outline represents a conceptual model at a basic level. So why didn't we just title this section "Develop a good outline?" Well, we could have, but that would've omitted an essential aspect of good instruction design: conceptualizing an overall user experience. This conceptualization connects with an existing user experience or imagines an enhanced user experience and allows the instruction designer to identify the content required to enable positive user interactions with the product. A conceptual model can also further clarify interrelationships between types of content and workflows, guiding the presentation of information in a way that a simple outline organized in a purely linear fashion would overlook.

Let's consider conceptual models for an instruction sheet that accompanies a lancing device. A lancing device pierces the skin to produce a drop of blood, which is then used to perform a blood glucose test or other similar diagnostic procedures. To test blood glucose, the user then touches a test strip to the droplet, places the test strip in a glucose meter, and ultimately receives a blood glucose reading after a few seconds.

Here are three, high-level outline options for the single-page instruction sheet:

Outline 1
- Product features
- Product purpose
- Obtaining a blood droplet
- Lancet disposal
- Alternate blood draw site selection

Outline 2
- Welcoming remarks
- Product purpose
- Warnings
- Preparing the lancet for use
- Lancing site selection
- Drawing blood
- Lancet disposal
- Frequently asked questions
- Manufacturer contact information

Outline 3
- Product description
- Safe use
- Lancing site selection
- Attaching the lancet
- Drawing blood
- Detaching the lancet
- Lancet disposal
- Operational tips
- Helpline number

Endowed with three conceptual models (i.e., outlines), you and other development team members could simply choose the one you think is best or iterate on all three to find one that you think would work. There's something to say for professional judgment.

However, ideally you would seek input from intended users, such as by means of interviews (see "Determine user needs" on page 40). For example, you could create some mockups that instantiate the different conceptual models so that people have a more complete basis for judgment. The mockups could include headings and "placeholder" content rather than be complete. Below, we share one example of such a mockup, noting that the headings represent actual sections but the content is placeholder text, the actual text for which might still be in development.

Lancing Device
Instruction Sheet

Device description
Lorem ipsum dolor sit amet, consectetur adipiscing elit. Lorem ipsum dolor sit amet, consectetur adipiscing elit.

Precautions
Warning: Lorem ipsum dolor sit amet, consectetur adipiscing elit, sed do eiusmod tempor incididunt ut labore et dolore magna aliqua.

Selecting a lancing site

Recommended sites:
- Lorem ipsum
- Dolor sit amet
- Consectetur

Unsafe sites:
- Lorem ipsum
- Dolor sit amet
- Consectetur

Preparing the lancing site

1. Lorem ipsum dolor
2. Lorem ipsum dolor
3. Lorem ipsum dolor
4. Lorem ipsum dolor

An example of a simple mockup, with placeholder text.

You can even get intended users involved earlier. In the case of an early user interview for a lancing device, you could demonstrate how the lancing device works in general and invite the user to participate in the conceptual modeling by performing an affinity diagramming exercise and arranging cards describing potential content into groupings that are logical for the user. This is called participatory design. You don't need a large number of users for this method to yield promising conceptual models; 5-8 users, interviewed together or individually, would do just fine.

So we've addressed why conceptual models are important and different ways to arrive at them, but one looming question remains: how many conceptual models should you consider? The answer is probably more than one but ultimately depends on the rigor of your conceptual design work. You could quickly converge on 2-3 models or evaluate a dozen or more. However, if you plan to collect user feedback, you might want to converge on 2-5 preferred models before soliciting feedback. After a judgment-based decision, perhaps informed by user feedback, the resulting preferred model might end up being one of the 2-5 original models or a hybrid of several with new additional elements (see "Explore multiple concepts" on page 63).

Create initial concepts

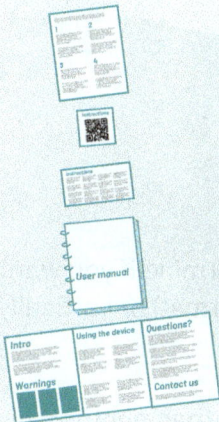

Discard unfeasible concepts

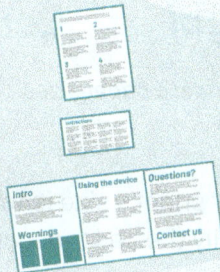

Identify preferred concepts

Conslidate & refine

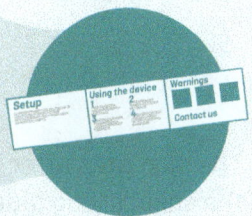

Multiple initial design concepts leading to a single, preferred concept.

27

Select orientation (aspect ratio)

From the tiny A10 to the grand B0, the International Organization for Standardization (ISO) has defined 33 standard paper sizes,[*] giving you a wide range of options to choose from — so long as you can accept a rectangular sheet.

Of course, most professional printers will let you print on custom shaped sheets, including square ones, for an added cost and longer production time. So, unless you have a good reason for printing instructions on a custom-cut square page, you'll likely end up using a rectangular page for your instructions.

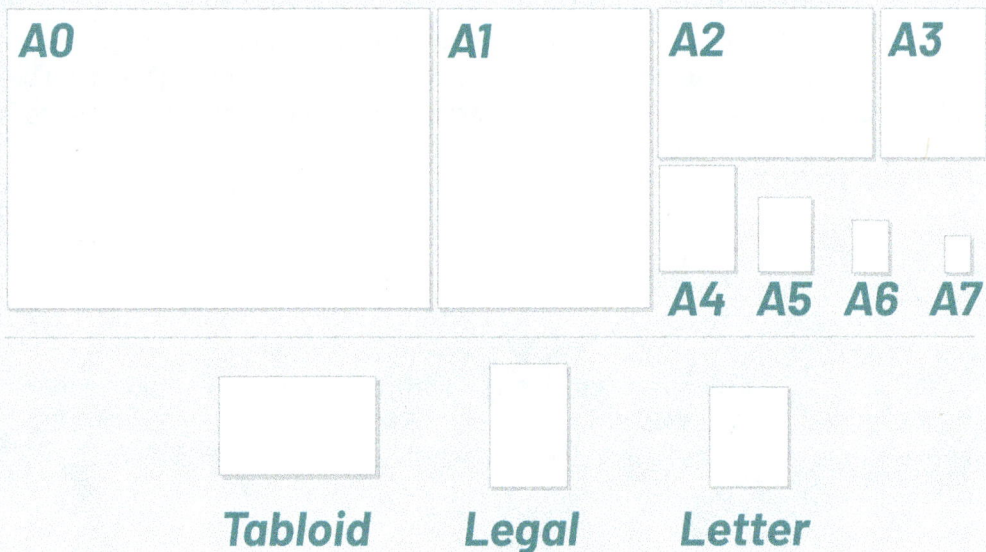

A sampling of standard paper sizes.

The next step after selecting your rectangular sheet's size is to determine if you want the sheet's long side to be oriented horizontally (i.e., a landscape format) or vertically (i.e., a portrait format).

We do not consider one page orientation to be dramatically better than another. It's usually best to explore both options and determine whether one orientation is better suited to your instructions' content and expected use. That said, each orientation can provide some benefits and shortcomings.

---

[*] International Organization for Standardization (ISO) - ISO 216:2007 - Writing paper and certain classes of printed matter — Trimmed sizes — A and B series, and indication of machine direction: https://www.iso.org/standard/36631.html

Portrait　　　**Landscape**

The primary advantage of a portrait orientation is that people are familiar with it from books. The orientation enables you to place text into shorter lines, which is helpful because the optimal line length for body text is 50-75 characters (including spaces).[*] Depending on the paper and font sizes, a portrait-orientated document might even accommodate two or more columns of text, making text line lengths nice and short. On the other hand, portrait orientations usually enable you to fit only a few frames (i.e., content cells) horizontally, thereby interrupting the visual flow from one frame to another.

The advantages and disadvantages of the landscape orientation are pretty much the inverse of the portrait orientation. Landscape orientations enable you to fit more frames in a single row, thereby providing extra continuity. However, body text may get too long if stretched across the entire page. You can work around this by placing body text in frames, which limits the number of characters and spaces per line. Another option is to place text in two or more columns, limiting the length and characters of each line. Of course, this may result in a somewhat awkward action of reading column one from top to bottom and then shifting your eyes back to the top to start the next column. Not an arduous visual task, but one that slows readers down increases the potential for reading errors.

Ultimately, you can make both orientations work well. However, a portrait orientation is typically better for predominantly text instructions, and a landscape orientation is more favorable for instructions that use a lot of graphics.

---

* Readability: The Optimal Line Length. Baymard Institute. https://baymard.com/blog/line-length-readability

What about digital displays?

Digital displays, such as those on personal computers and mobile devices (i.e., phones and tablets), are no different than paper when it comes to rectangularity, with the most common aspect ratios (width:height) being 4:3 and 16:9. That said, there are a couple distinct characteristics that might affect how instructions are viewed on digital displays:

- Slight differences in aspect ratio or resolution among different displays could result in the instructions appearing differently on different devices (e.g., on an Apple iPhone vs. an Android phone). As a best practice, you'll want to test your instructions on a few different devices and displays to ensure that they accommodate their shape and aspect ratio.

- Mobile device users have the option of switching their device's orientation between landscape and portrait. You may wish to explore the option of presenting your instructions using a so-called responsive layout, which dynamically rearranges the instructions' content to best suit the current orientation.

SIDEBAR
How many standard paper sizes are there?

Different countries and regions have adopted various systems for standard paper sizes, so it's difficult to estimate the total number of standard paper sizes in the world. That said, the two most common systems are:

- The ISO 216 system, which is used by most countries and includes three series (A, B, and C series). Each series includes multiple sizes (A0-A10, B0-B10, and C0-C10), with the A series being most widely available among printers and the A4 being the most common size for everyday printing.

- The North American system, which includes Letter, Legal, Ledger/Tabloid, and other sizes defined by ANSI (American National Standards Institute).

Each system has numerous sizes, and when you consider additional systems or regional variations, there are over a hundred standard paper sizes in total.

28

Facilitate translation

In "Accommodate diverse audiences" on page 91 we share our thoughts about designing instructions for a diverse audience, including people from different countries. However, there will be times when you will want to create localized versions of the instructions if the associated medical device will be marketed internationally.

Unless your instructions can be solely graphical — a difficult challenge — you may need to create them in different languages to accommodate people in various countries or even people within a single country. And no, we don't think that IKEA's wordless approach to instructions will work for most medical products, even though their illustrations are sometimes impressively clear.

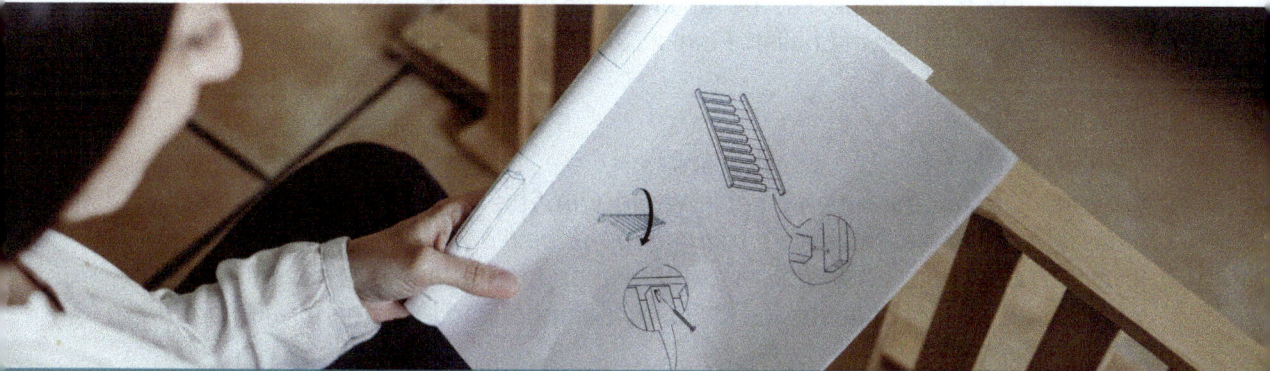

IKEA's wordless instructions often feature clever depictions of complex actions, though they've been known to induce the occasional use error.

We suggest you consider the following factors as you develop instructions that must be translated into one or more languages:

- When there is enough space, present instructions in multiple languages side-by-side, making sure that users can visually distinguish them to avoid confusion.

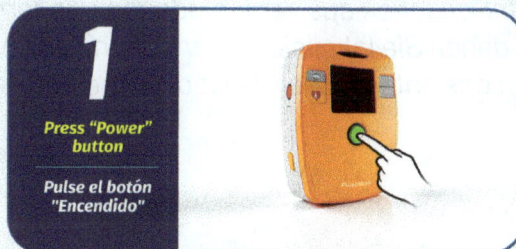

An example instruction that features a Spanish prompt beneath the English one.

- For single-sided, single-sheet instructions, print the instructions with one language on one side and another language on the second side.

- For multi-page instructions, make the first section one language (e.g., English) and the second section another language (e.g., Spanish).

- When arranging content on the page, make sure to consider the amount of space that might be required for translated text. Reportedly, "Finnish gets the content across with the least [number] of words but with the longest word length, averaging 7.899 letters... In contrast, Italian, French and Spanish use more words to express a meaning. The individual words in our sample text are in fact quite short compared to the other languages, with an average total of 5.462 to 5.675 letters."* Anecdotally, we have found that an English-to-German translation results in the text consuming 30%-40% more space. For the fun of it, here is a comparison of this section's first paragraph written in English, Finnish, and German, thanks to Google Translate.

Finnish (34 words, 282 characters)
Kohtaan "Ota huomioon erilaiset käyttäjäryhmät" sivulla 103 jaamme ajatuksiamme ohjeiden suunnittelusta moninaisille yleisöille, mukaan lukien eri maista tulevat ihmiset. Kuitenkin on tilanteita, joissa haluat luoda lokalisoidut versiot ohjeista, jos kyseistä lääkinnällistä laitetta markkinoidaan kansainvälisesti.

English (48 words, 281 characters)
In "Accommodate diverse audiences" on page 91 we share our thoughts about designing instructions for a diverse audience, including people from different countries. However, there will be times when you will want to create localized versions of the instructions if the associated medical device will be marketed internationally.

German (46 words, 322 characters)
In "Unterschiedliche Zielgruppen berücksichtigen" auf Seite 103 teilen wir unsere Gedanken über die Gestaltung von Anleitungen für ein vielfältiges Publikum, einschließlich Menschen aus verschiedenen Ländern. Es wird jedoch Zeiten geben, in denen Sie lokalisierte Versionen der Anleitungen erstellen möchten, wenn das entsprechende Medizinprodukt international vermarktet wird.

---

* Why does content expand when translated? Intercontact translations. 08 January 2020. https://www.inter-contact.de/en/blog/text-length-languages

- Consider the potential need to adjust the text size and formatting of translated text to make sure it's legible. Languages that use non-Latin characters, such as Arabic and Chinese, often require larger text and greater line heights to ensure legibility.*†

Here are some more tips pertaining to the writing itself that will ease translation, adapted from *10 Essential Tips for Writing for Translation*.‡

- Write in the active voice (a tip you'll see in other sections as well) because it is easier to translate correctly

- Keep sentences short

- Avoid noun strings (e.g., robot user interface control panel)

- Avoid abbreviations, acronyms, and initialisms unless (1) they are in practically universal use and (2) spelling things out would be more confusing

- Avoid the use of humor, idioms, jargon, and metaphors

Also, beware of creating graphics with embedded text that cannot be directly edited by translators. Similarly, numbers and math symbols might also need to be "translated" to languages that don't use Arabic numerals, such as Arabic (which uses Hindu-Arabic numerals) and Thai (which uses Thai numerals).

The graphic on the left includes only a number, which would not require translation for most languages, while the graphic on the right includes text that would require translation.

---

* Effects of Font Size, Stroke Width, and Character Complexity on the Legibility of Chinese Characters https://onlinelibrary.wiley.com/doi/abs/10.1002/hfm.20663

† Towards Usability Guidelines for the Design of Effective Arabic Websites https://arxiv.org/ftp/arxiv/papers/2011/2011.02933.pdf#:~:text=For%20the%20English%20language%2C%20font,at%20font%20size%2014%20points

‡ 10 Essential Tips for Writing for Translation. Lionbridge. https://www.lionbridge.com/blog/translation-localization/writing-for-translation-10-expert-tips-to-boost-content-quality/

29
Take it step-by-step

Most people don't like being told what to do. It starts when people are toddlers and continues through adulthood, right? Most of us are familiar with the stereotypical temper tantrum of young children, the rebellious teenage phase, and even the headstrong adult insisting they know best.

But, when it comes to instructions, telling people what to do is a must, and they appreciate it. More specifically, people like to be led, step-by-step, through procedures.

This will sound ridiculously obvious, but one of the best things you can do to keep people on the right track is to number steps in instructions (i.e., Step 1, Step 2...Step N).

And there is no point in being visually subtle. Make sure the step numbers are visually conspicuous. This can be accomplished by making the step numbers hanging headers, providing generous blank space between steps, and even using the word "Step" before the numeral.

Labeling and differentiating substeps, such as by indentation and hierarchical naming (e.g., Step 1.1, Step 1 (a)), can also specify and clarify the correct procedural order.

Things get more complicated when a procedure is nonlinear, presenting branching possibilities. In other words, when a procedure includes conditional steps that users might need to perform in some situations but not others. Even these possibilities can be clearly explained by using the words *if* and *then*, or *else* (i.e., *otherwise*).

Initial setup

1 **Unpack device**
Remove the VitalCheck Monitor and all accessories from the packaging.

2 **Install batteries**
Open the battery compartment and install two (2) AA batteries (included in the box).

3 **Check fluid**
Inspect the fluid window to confirm that the device has at least 10mL of fluid.

4 **Power on device**
Press and release the power button.

Prominent step numbers clearly convey the order of actions users must take.

30
Accommodate diverse audiences

Culture, race, religion, age, sex, gender, sexual orientation, cognitive, and physical abilities[*] are the common categories people tend to associate with diversity. It's important to consider these factors when creating instructions, but there are additional forms of diversity to consider, including users' language skills, reading level, active vocabulary, and learning style.

Imagine you are designing instructions for nurses to operate an advanced hospital bed. You might consider this user population as relatively homogeneous. But don't assume too much. A given nurse may be a certified nurse assistant, licensed practical nurse, licensed vocational nurse, registered nurse, nurse midwife, nurse anesthetist, or nurse practitioner.[†] This leaves plenty of room for diversity.

A seemingly homogeneous user group, such as nurses, can include significant diversity in terms of skill, education, and responsibilities.

Therefore, even if your user population is limited to nurses, for example, you are targeting a very diverse population. Nurses are quite diverse in all the ways listed above, and in other ways as well. Sure, you assume — correctly or not — there will be differences among nurses practicing in Germany, the Philippines, or the United States, for example. But there could be more profound differences between two nurses born in the same country who went to the same nursing school.

---

* What are the types of diversity? https://resources.workable.com/hr-terms/the-types-of-diversity
† Jennifer Chi. Careers for nurses: Opportunities and options. U.S. Bureau of Labor and Statistics. https://www.bls.gov/careeroutlook/2020/article/careers-for-nurses-opportunities-and-options.htm

The people who use your instructions are likely to be quite diverse.

Here are some ways, adapted in part from *Writing for Diverse Audiences*,[*] to develop instructions for a diverse user population:

- Avoid idioms, such as "good rule of thumb" because people for whom English is a second language might be unfamiliar with such expressions.

- Use simpler verbs, such as "use" instead of "utilize."

- Avoid double negatives, such as "not uncommon."

- Write in the active voice (e.g., "Wash your hands before injecting to reduce the chance of infection"), as we explain in "Use active voice construction" on page 150.

- Avoid specifying gender. Often, you can do this by referring to people by their functional title (physician, nurse, patient) rather than their gender (her, his). You can also refer to people using the gender-neutral pronouns, such as "their" or "they." Note that the "grammar police" used to frown on the use of "they" and "their" when referring to a single person, but this usage has become common in conjunction with the embrace of diversity, equity, and inclusion goals.

✖	✔
Gendered	**Gender-neutral**
Once the doctor has made **his** diagnosis, **he** should send it to the nurse so **she** can inform the patient.	Once the doctor has made **their** diagnosis, **they** should send it to the nurse **who** can inform the patient.

---

* Writing for Diverse Audiences. Teaching and Learning Inquiry. https://journalhosting.ucalgary.ca/index.php/TLI/writing-for-diversity

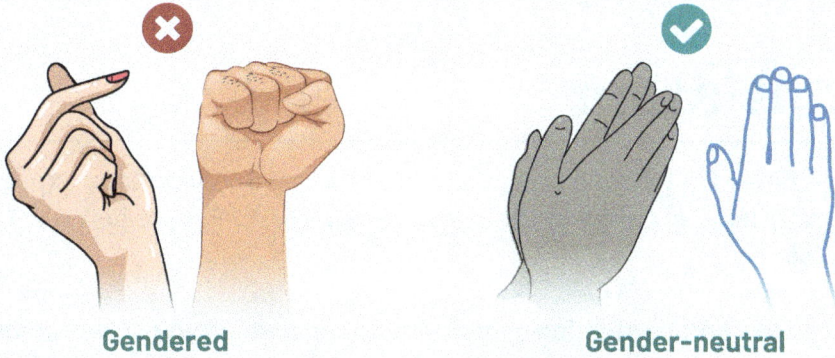

Gendered   **Gender-neutral**

*Many body parts can also be depicted in a gender-neutral
way to make them more relatable to users.*

- Depict different types of people representing the intended user population, rather than just one particular type. Also, when illustrating body parts, aim to depict them in a gender-neutral manner. That way, more readers will find it relatable.

- Avoid defining people based on their medical conditions. For example, refer to a "person with diabetes" rather than a "diabetic," a "person with a visual impairment" rather than a "blind person," or "a person with a disability" instead of a "handicapped person."

- Do not hesitate to write about skin color when it is medically relevant, such as when indicating if a dermatological laser device can be used safely and effectively on skin of a certain color. Hair growth retarding devices that use intense pulsed light, for example, are contraindicated for people with dark brown skin because the high concentration of melanin in their skin causes it to absorb the laser light and result in burns.

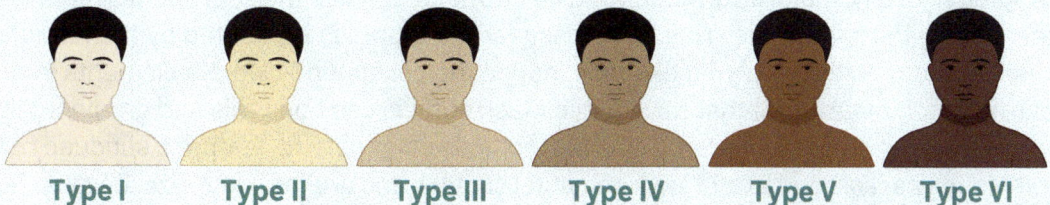

Type I   **Type II**   **Type III**   **Type IV**   **Type V**   **Type VI**

*This graphic conveys the six skin tones defined in the Fitzpatrick scale,
which classifies skin by sun reaction and tanning ability. It is also a good
example of depicting facial features in a race-neutral way.*

31

Be consistent

Instructions differ from books and magazine articles in that nobody expects them to break new ground in terms of their graphics or prose. People just want them to be easy to follow.

This is achieved in part by meeting users' expectations, which you do by conforming to convention and, perhaps more importantly, being internally consistent. Inconsistency in most forms will confuse users and possibly induce mistakes. It can also be distracting and reduce the appeal or perceived authority of the document.

Below, we describe some of the important ways in which instructions should be consistent:

Sentence structure. Minimize arbitrary variations in sentence structure, such as sometimes writing in the active versus passive voice (see "Use active voice construction" on page 150) or sometimes starting an instruction with an imperative phrase and sometimes not (see "Permissive versus prohibitive phrasing" on page 152).

Wording. Avoid using synonyms for the sake of variety. For example, do not switch arbitrarily between the words "use" and "utilize." In particular, refer to product components using consistent terms (e.g., always using the term "cartridge," rather than mixing "cartridge," "pack," and "unit").

Headings. Ascribe to a uniform information hierarchy, reinforced by headings that have a consistent appearance including their size, color, placement, and spacing relative to adjacent content.

Margins. Establish standard margins for specific types of content to create visually pleasing separation and avoid the careless appearance that results from widely varying margins.

Graphics. Ideally, make all graphics and illustrations look like the same person created them using the same style. This is the case for the graphics presented by the website wikiHow.com. In the following example, notice the consistently simple elements (e.g., pen-injector, mask, AED) that have a thin black outline, use color fills and gradient fills to give objects greater form, employ a harmonious color palette, feature a subdued but interesting background to add environmental context, and are the same size. Developing a graphic design style guide can be helpful.

The website wikiHow.com features consistently styled illustrations related to over 2,000 topics. (images courtesy of WikiHow, Inc.)

Color. Use color consistently to communicate extra meaning (e.g., critical warning) as opposed to decoration (see "Use color effectively" on page 100).

Units. It is particularly important to use consistent units of measurement. Do not switch between imperial (English) units and metric units because it will induce use errors. The universal solution is to use metric units and put the imperial units in parenthesis. The exception might be if you are writing instructions for laypeople in the US who are more accustomed to imperial units (US clinicians, however, are generally accustomed to the metric system).

Weight: 70 kg (155 lbs)

Demarcation. Decide if you will use outlines or color to encapsulate content. Avoid mixing multiple approaches unless you want to sharply differentiate content. For example, use bold lines to demarcate the steps involved in using the medical product, but use a background color fill to demarcate the manufacturer's contact information (e.g., a helpline phone number).

Alignment. Apply the same justification to all text of a particular type (i.e., headings, body text, captions, etc.) in your document. To improve legibility, we generally suggest left-justifying text and using a ragged right margin. This allows for normal word spacing and more natural reading as compared to the variable word spacing that occurs with full justification.

Balance the composition. Try to distribute information evenly across an instruction sheet or page. Avoid high information density in one part that makes it challenging to read while leaving others portions sparse.

Comparison of balanced (left) and imbalanced (right) compositions.

Maintaining consistency might seem like a straightforward goal but achieving it can be a challenge. This is particularly true when multiple people are collaborating on a single document or complementary instructional materials (e.g., a user manual, quick reference guide, and instructional video).

One way to ensure consistency is to create a style guide — a document that provides rules, guidelines, and suggestions for maintaining consistency in the instructions' written and visual content. Style guides typically include subsections for each key aspect of the instructions, such as one for the fonts and layout, one for graphics, and one for copy. You can also share style guides across projects and teams to maintain consistency across instructions for different products produced by the same organization.

32

Follow conventions

Sometimes, being unconventional is a good thing. In the fashion world, for example, it suggests that you are creative, an original thinker, avant-garde.

In art, literature, and music, avant-garde refers to works that break conventions, introducing or experimenting with novel forms, often in a subversive manner. Avant-garde is originally a French term, meaning in English vanguard or advance guard (the part of an army that goes forward ahead of the rest).[*]

We say vive l'avant-garde! (This is the second time we've used French in this book.) But is there a place for the unconventional when it comes to medical device instructions?

Examples of unconventional product design: Espresso machine by Carlo Borer (upper left) and bicycle design by Robert Egger (lower right).

---

[*] Tate Art Term: Avant-garde. https://www.tate.org.uk/art/art-terms/a/avant-garde

If the instructions are excellent because they don't conform with common, mediocre examples, then our answer is yes. But, if instructions break conventions that would otherwise make them more usable, the answer is a bold no. For this reason, we suggest shying away from experimentations such as:

- Eliminating almost all words

- Using a cartoon character to graphically demonstrate device use

- Introducing humor in the same manner of some airline safety videos

- Organizing content in a spiral

- Utilizing a non-linear organization that allows users to read the instructions in their preferred order

- Using experimental typography that makes the instructions have a distinct brand

- Using abstract art to represent the device's operating principles

- Using slang and informal language

All of these introduce the risk of poor communication with users and possibly making the instructions conflict with regulatory requirements.

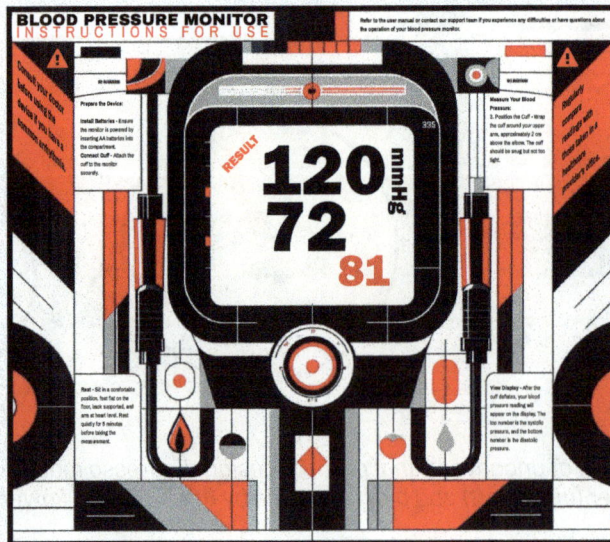

These constructivism-inspired instructions for using a blood pressure monitor might catch some users' attention but are unlikely to help them operate their device effectively.

Here is a small sample of conventions you should probably follow, many of which are mentioned elsewhere in the book:

- Order content to match the user population's normal reading pattern, which is right-to-left and top-to-bottom in many cultures

- Make important information more visually conspicuous than less important information

- Use sentence capitalization for body text

- Use arrows to indicate direction and point to specific items

- Place blank space between text and graphics

- Use dotted lines to show how product components align

- Number procedural steps. Using letters (a, b, c, d...) is less conventional but also okay

- Use the safety alert symbol to draw attention to warnings

- Demarcate functionally related content from other content using blank space or demarcating features, such as a line

- Place captions below the related illustration or table

- Align content to a grid

- Strive to give a page of content a balanced appearance

- Left-justify body text

However, if you do innovate in ways that defy convention and develop avant-garde instructions, be sure to conduct a lot of user comprehension and acceptance testing. Plus, make sure you are not in violation of published regulatory expectations.

33

Use color effectively

You could fill an entire book with advice on how to use color effectively, and some authors have done so. In this section, we'll focus on two aspects of color:

- Using color to frame information
- Using color to differentiate information

Framing

We use the term "frame" in the conventional way, as in framing a photograph. Frames establish boundaries and contain content. Frames help focus attention. They provide demarcation, such as between functionally related groups of information.

Within instructions, you can use a colored demarcation line to frame a set of content. In some cases, it works even better to place content on a colored background, which we also consider to be framing. You should take care when choosing the outline or background color. As discussed in "Instructions should harmonize with the product" on page 17, the color can be used to reinforce a brand. The chosen color might convey meaning, such as using a red frame around a warning. Also, the color choice might reinforce a hierarchy, such as using dark and light blues to frame primary and secondary information, respectively.

You can achieve visually appealing results when you use a gradient fill. For example, you can consistently use background frames that start with a selected hue at the top and then fade somewhat or completely toward the bottom of the frame.

Using your peak flow meter

1
Prepare meter
Make sure meter is clean and dry.

2
Take deep breath
Inhale deeply, filling your lungs completely.

3
Position meter
Place mouthpiece in mouth and tighten lips around it.

4
Blow out
Blow out as hard and as fast as you can in a single blow.

These instructions use a background gradient to frame each use step and help differentiate the title from the contents.

Another interesting effect is to put a transparent frame over a background image so that overlying text draws attention and is sufficiently legible. However, this effect is contraindicated when legibility is critical, which is when you should steer toward black letters on a white background (or the reverse) to achieve maximum contrast.

Differentiation

The common sense use of color is to differentiate content. So, consider how to use color as effectively as possible to provide instructions. While the opportunities and constraints will differ from designing a software user interface, the principle is the same.

The simplified patient monitor screen below differentiates content with color very well, helping to associate waveforms and numeric values. Moreover, the color assignments align with patient monitor conventions and user expectations, at least for a parameter such as arterial blood pressure, where there is an association of blood and the color red.

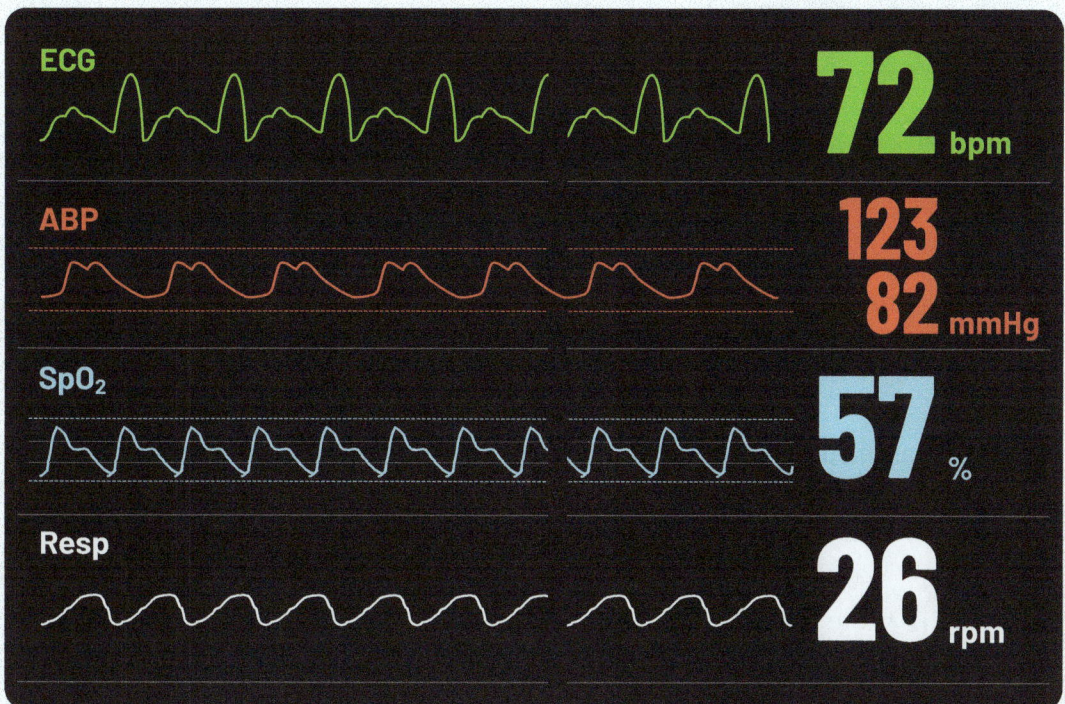

Alarms and warnings certainly warrant color coding that matches the associated hierarchies. In the case of warnings, the hierarchy includes red, amber/orange, yellow, and blue (see "Warn users about potential risks" on page 106).

ANSI Safety Red
PMS 186C / Hex code #bd2024

ANSI Safety Orange
PMS 151C / Hex code #ff7900

ANSI Safety Yellow
PMS 109C / Hex code #ffe100

ANSI Safety Blue
PMS 285C / Hex code #004488

Be mindful that color can convey intended and unintended meanings. A lot depends on a user's culture, the colors used in standardized ways in their line of work, and individual experience. In our daily lives, using a particular color might be associated with the following meanings.

RED	ORANGE	YELLOW	GREEN	BLUE
• Anger • Danger • Energy • Heat • Love • Passion • Power • Strength • Warning	• Confidence • Encouragement • Excitement • Extroversion • Health • Vitality	• Brightness • Energy • Happiness • Hope • Intellect • Sunshine • Vibrancy • Youth	• Earth • Envy • Freshness • Growth • Guilt • Harmony • Jealousy • Money • Nature	• Cold • Fear • Intelligence • Loyalty • Masculinity • Peace • Security • Tranquility • Trust

PURPLE	PINK	BROWN	BLACK	WHITE
• Ambition • Fantasy • Luxury • Moodiness • Mystery • Nobility • Royalty • Spirituality	• Compassion • Femininity • Happiness • Hope • Immaturity • Inspiration • Playfulness • Sweetness	• Conservative • Earth • Honesty • Longevity • Nature • Outdoors • Reliability	• Classiness • Death • Elegance • Formality • Luxury • Mystery • Power • Protection	• Cleanliness • Ease • Freshness • Innocence • Purity • Simplicity

In the medical and pharmaceutical industries, some colors have been standardized. For example, there are standard uses to label medical gasses.

	US Color Code		ISO Color Code	
Carbon Dioxide	Grey		Grey	
He-O2	Brown	Green	Brown	White
Medical Air	Yellow		Black	White
Nitrogen	Black		Black	
Nitrous Oxide	Blue		Blue	
O2-He	Green	Brown	White	Brown
Oxygen	Green		White	

In hospitals, specific colors are used to alert staff to specific conditions and events. Patient wristbands feature color coding, too.

Hospital event codes
Bomb threat
Disaster
Fire
Hostage taking
In facility hazardous spill
Medical emergency
Medical emergency – infant/child
Missing person
Precautionary evacuation
Violent/behavioral situation

Patient hand bands
Allergy
Do Not Resuscitate (DNR)
Fall risk
Fluid restriction
Latex allergy
Restricted extremity

It's good to know about these standard uses and others. It does not rule out using a particular color for a particular purpose within instructions, but consider how some people might react intellectually and emotionally to a given color.

We'll wrap up this section with a few rules of thumb regarding color use.

- Try to limit the use of color and treat it as a complement to a document that is otherwise monochromatic (e.g., grayscale) or purely black and white. Many designers use different values (i.e., multiple shades) of one color (a single hue) as a baseline for framing and headings and then use other colors to convey specific meaning and make content conspicuous.

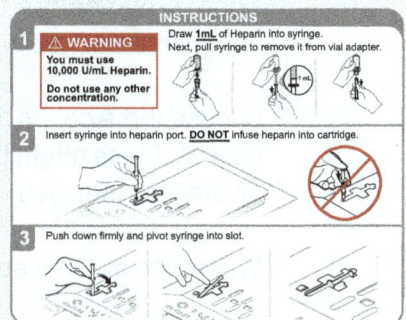

- Make sure that the instructions remain legible and informative to someone who has color blindness. One trick is to view or print the instructions after digitally removing the color (i.e., desaturating), which leaves white, grays, and black. Another is to initially design the instructions without color and then add color strategically and sparingly. In the case of effective color use, less is usually more.

These instructions are effective without color (left) but benefit from selective use of color to highlight key details (right).

- An overall harmonious appearance comes from using harmonious colors. You can achieve this through various means, including the following:

  · Choose colors with equivalent degrees of saturation (color intensity versus grayness) and value (lightness versus darkness).

  · Choose two colors that are on opposite sides of the color wheel (i.e., complementary colors).

  · Choose three colors that are equidistant on the color wheel (i.e., analogous colors). These might be concentrated on one portion of the wheel (e.g., over an arc less than 360 degrees) or the full wheel (360 degrees).

  · Choose colors that appear in a natural scene, such as a desert or meadow.

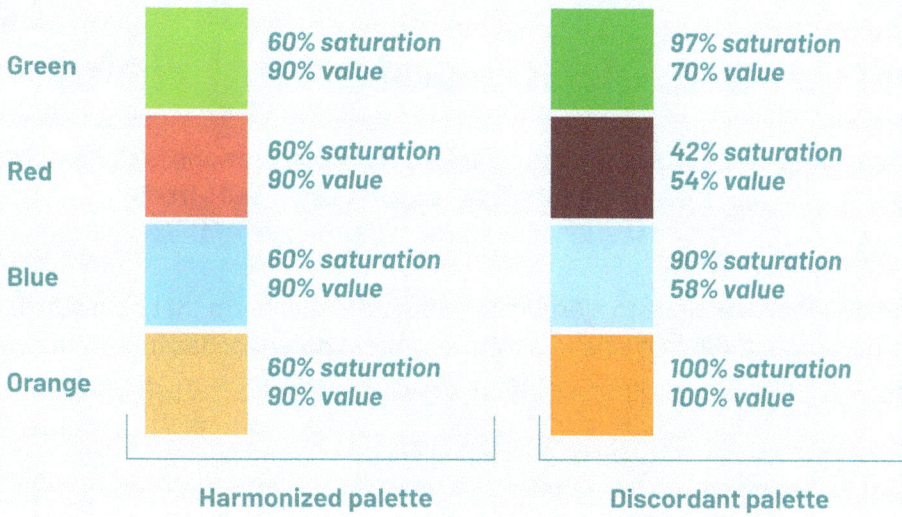

Green		60% saturation 90% value	97% saturation 70% value
Red		60% saturation 90% value	42% saturation 54% value
Blue		60% saturation 90% value	90% saturation 58% value
Orange		60% saturation 90% value	100% saturation 100% value

Harmonized palette   Discordant palette

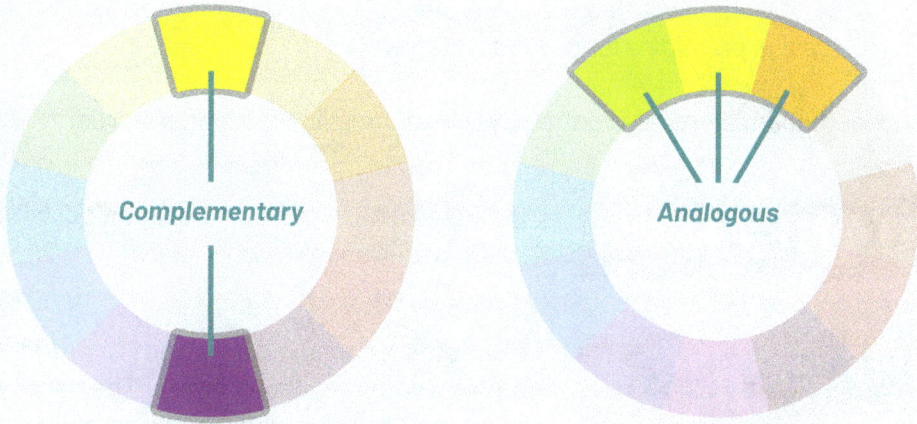

Complementary   *Analogous*

34

Warn users about potential risks

A distinguishing characteristic of medical product instructions is the emphasis on safe and effective use. An important part of ensuring safe and effective use is making users aware of potential hazards associated with the use of the product at hand.

It's common for instructions to alert users about potential harm that could occur when operating a product. When trying to alert users about potential hazards, it's a good idea to follow conventions such as those identified in two relevant standards:

- ANSI Z535 - which is a set of six standards that define best practices for safety-related alerts, with standards Z535.1 and Z535.3 focusing on alert colors and symbols, respectively

- ISO 3864-2 - which is titled *Graphical symbols — Safety colors and safety signs — Part 2: Design principles for product safety labels*

Standard-compliant alerts are topped by one of the following signal words:

⚠ **DANGER**	Indicates a hazard that, if not avoided, will cause death or serious injury.
⚠ **WARNING** / **WARNING**	Indicates a hazard that, if not avoided, could cause death or serious injury. The same signal word may be used, without the triangular safety symbol, to protect against the chance of serious and moderate property damage.
⚠ **CAUTION** / **CAUTION**	Indicates a hazard that, if not avoided, could cause moderate or serious injury. The same signal word may be used, without the triangular safety symbol and optional text-ground inverted, to protect against the chance of moderate and minor property damage.
NOTICE	Does not pertain to the threat of operator injury. Instead, it is used to inform the user about something considered important to effective use. Note that the triangular safety symbol is never paired with the signal word NOTICE.

The triangular safety symbols vary slightly in design depending on whether you are following the ANSI Z535 standard alone or complying with both ANSI Z535 and ISO 3864-2. One triangular safety symbol design is not definitively better than the other.

ANSI Z535

ISO 3864-2

Don't worry if you see plenty of instructions with alerts that do not conform to the guidance above. Compliance with the guidance is spotty. It's common for some designers and attorneys to assert their judgment about what is best versus following the standard. So, you will see such cases as using the signal word WARNING when exposure to a hazard will not necessarily result in death or serious injury.

The text message accompanying the signal word needs to be short and to the point. You can indicate what not to do to avoid a hazard; in other words, prohibit unsafe behavior. Alternatively, you can indicate what to do to avoid a hazard; in other words, encourage safe behavior.

Consider the following examples, which show original drafts of alert messages alongside revised alerts that are more compliant with the standards:

Original draft	Standard-compliant revision	Explanation
CAUTION: Be careful when using needles	CAUTION: Sharp needle – Use with care	The improved example specifies the hazard (sharp needle) and the action to avoid it (use with care).
WARNING: Don't let the syringe fall	WARNING: Syringe contamination risk – Handle securely	The improved example identifies the hazard (syringe contamination) and the recommended action (handling the syringe securely).
NOTICE: Proper storage is important for medicines	NOTICE: Store medicines in a cool, dry place	The improved example is concise and clearly instructs the user on the necessary storage conditions (a cool and dry location).

In instructions, you should place alerts where they will be read and followed at the right time during a task. For example, warn a user about the chance of a needlestick injury before directing them to uncap a needle on an injector.

Differentiate the alert from other text to help it get noticed. One way to do this is to print the signal word in its intended color (e.g., red, orange, yellow, or blue). Then, use the same color to make the text stand out as well, such as printing the text on a very light background of the same hue, perhaps with a darker shade or black outline.

You might want to complement the alert text with a graphic, such as with warning signs.

⚠WARNING

Biohazard materials
Risk of infection.
Use appropriate protective equipment and follow safety procedures.

⚠CAUTION

Flying debris
Risk of eye injury.
Wear safety goggles or face shield at all times.

⚠DANGER

High voltage.

Contact will cause shock burns or death.

Disconnect electrical supply before servicing.

Examples of ANSI Z535 compliant warning signs.

35
Help users resolve problems

When are users most likely to seek instructions? Not only when they are learning to use a product, but also when they face a problem. For example, they might be dealing with an unfamiliar alarm or recovering from a use error, unsure how to proceed. Let's talk about the latter case.

Instructions in various forms occasionally include a troubleshooting section, which may be a dedicated page or a portion of an instruction sheet. In our experience watching participants in usability tests deal with simulated problems, people go straight to the troubleshooting section to see what they should do to fix things. Many users expect instructions to include this content and benefit from it.

Medical product manufacturers often present troubleshooting content in lists or tables. In a table, for example, you might see one column describing potential problems and another the associated solutions. The key is for them to be written clearly and concisely and perhaps be complemented by graphics. Category headings over a set of related rows in the table also help with navigation (e.g., the heading "Power" in conjunction with troubleshooting tips pertaining to a loss of power due to various causes).

In some cases, a troubleshooting flowchart might be the best way to guide users on how to clarify and then solve a problem. Such charts visually depict the problem-solving process in a step-by-step manner. They are especially helpful for resolving more complex or multi-step problems, as they present the troubleshooting process in a series of actionable steps. They can also be more helpful than tables or lists when the troubleshooting process involves decision points, conditional steps, or multiple possible paths to a solution.

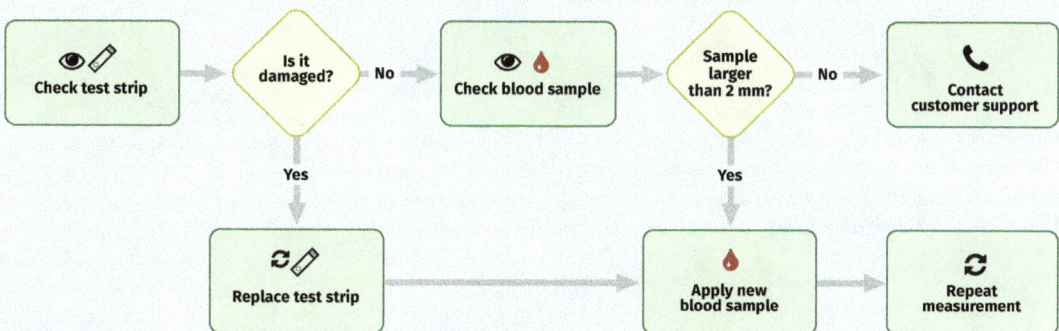

A sample flowchart explaining how to troubleshoot a glucose meter's "Insufficient Blood Sample" error.

Instead of placing troubleshooting content in a dedicated section, you may choose to embed troubleshooting tips within the primary content (e.g., task instruction). In "Warn users about potential risks" on page 106, we advise doing the same with warnings.

However, placing troubleshooting content alongside task instruction could create visual congestion and overload users with information. Also, this approach might help resolve problems that occur while performing a task, but not problems that arise when the user is focused on a different task or when not even interacting with the product.

Regardless of its location in your instructions, troubleshooting content should get to the point of how to correct a problem. It is neither the place to provide basic training, nor go into depth on a topic that is largely irrelevant to fixing the problem at hand. You want to help the user fix a problem as quickly as possible. In some cases, you might give users options on how to fix problems if there are tradeoffs.

Like troubleshooting content, information about alarm conditions is often presented in a consolidated list or table. Some medical devices present errors in a rather cryptic, coded way (e.g., E7), perhaps because of severe display limitations. A better option might be to show a screenshot of the alarm or to describe the alarm condition in some detail. Then, as with troubleshooting tips, the goal is to tell the user how to deal with the problem as quickly and effectively as possible. Be clear and concise. Also be directive, meaning tell the person exactly what to do (see "Use good grammar when directing users" on page 136).

Sometimes, even well-written troubleshooting instructions don't suffice. For this reason, it's a good practice to end the troubleshooting section with options for follow-up support, which might include phone, chat, or email contact details for customer support.

36
Use composition principles

Many types of creative people create compositions, including songs, essays, and art. In the context of this book, we use the term in a similar manner to a newspaper editor, referring to the arrangement of content on a page.

Good page composition is usually easy to differentiate from bad. It's apparent at-a-glance. Something looks right versus wrong. You don't need to know the formal rules of composition to see the difference. However, knowing the rules can help produce a composition that is likely to please most people and, more importantly, help people acquire information more quickly and completely. To do this, you arrange and size information in an advantageous manner. Here are some suggestions for producing pleasing compositions:

Content blocks. Think of content, such as a paragraph of text or an illustration with its caption as a "block." You can create visual balance by distributing the blocks evenly both vertically and horizontally across a page.

Balance. Create visual balance among blocks of content by distributing them fairly evenly on the page, including horizontally, vertically, and diagonally. An abundance of imbalance could lead the viewer to overlook certain blocks.

VitalCheck Home Health Monitor

Content block

Introduction

VitalCheck Home Health Monitor is a comprehensive home-use medical device that measures blood pressure, pulse rate, blood oxygen level, body temperature, and provides a basic EKG reading.

Initial setup

1 **Unpack device**
Remove the VitalCheck Monitor and all accessories from the packaging.

2 **Install batteries**
Open the battery compartment and install two (2) AA batteries (included in the box).

✓ *Balanced composition*

VitalCheck Home Health Monitor

Introduction

VitalCheck Home Health Monitor is a comprehensive home-use medical device that measures blood pressure, pulse rate, blood oxygen level, body temperature, and provides a basic EKG reading.

Initial setup

1 **Unpack device**
Remove the VitalCheck Monitor and all accessories from the packaging.

2 **Install batteries**
Open the battery compartment and install two (2) AA batteries (included in the box).

✗ *Imbalanced composition*

Grid. People usually prefer things aligned to a grid. The things could be chairs on an auditorium floor, groceries on shelves, or content on a page. Therefore, establish a page grid just as one might when composing a presentation slide deck. Then, size the content blocks to match the size of the cells, and "snap" them to the grid. This approach keeps blocks from stepping over visual boundaries, which can "annoy" the eye. It also leads to blocks that might be 1, 2, or 3 cells wide, for example, while eschewing blocks that are 1.7, 2.3, and 3.6 cells wide.

Two-column grid   *Three-column grid*   *Four-column grid*

Consistency. Make similar types of content look similar so that people do not try to derive meaning from visual differences. For example, printing one photo in color and another in monochrome makes little sense unless there was a purpose in doing so. On a related note, content looks disorganized and is more difficult for readers to follow when it is not consistently left-, center-, or right-justified (i.e., aligned).

Aspect ratio. Both pages and blocks of content may be taller than they are wide or vice versa. Many designers believe that viewers prefer the symmetry of squares and rectangles that match the golden ratio, which is 1:1.618 (landscape or portrait orientation).

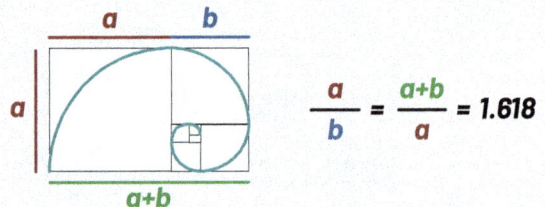

$$\frac{a}{b} = \frac{a+b}{a} = 1.618$$

An illustration of the golden ratio.

Order. In most languages, people read left-to-right and top-to-bottom. Take this into consideration when arranging content so that people view it in the intended order. The following illustration shows how people tend to move their eye gaze across a page, and the importance people assign to information in each area. Expect users to scan the page in a similar manner, but remember that you can influence their gaze using navigation aids such as arrows and frame/step numbers. "The Gutenberg Rule is used to show a user behavior known as reading gravity, the western habit of reading left-to-right, top-to-bottom."*

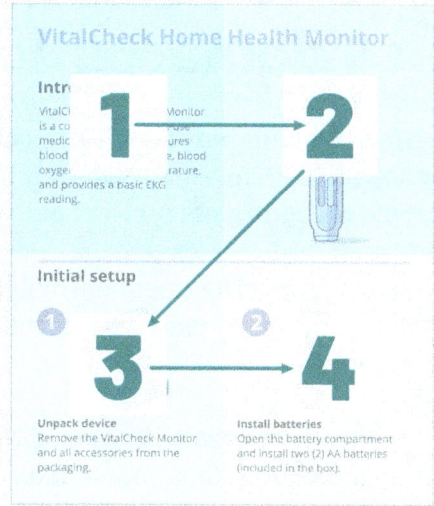

Break the rules. Equipped with an understanding of composition rules, feel free to break them not due to negligence but with purpose. Indeed, a block of content that does not conform to an implicit grid or throws off a page's balance could work to draw attention to critical content. Note that exceptions to the rules are usually the most effective when they stand out from adjacent content that does adhere to convention.

Typical reading order in left-to-right languages.

SIDEBAR
Composition for right-to-left languages

Although most languages are read left-to-right, some languages, such as Arabic, Farsi, and Hebrew, are read from right-to-left. When creating layouts for so-called RTL (right-to-left) languages, the best practice is usually to mirror the layout used for left-to-right languages and to utilize right-aligned text primarily, as opposed to left-aligned text.

Notably, numbers in RTL languages are still written left-to-right, which can affect the layout of items such as lists and tables that include a mix of alphabetical and numeric content.

English (top) and Hebrew (bottom) versions of the same instructions.

---

* Mário R. Andrade. Gutenberg Diagram — Why you should know it and use it. Jun 9, 2013. https://medium.com/user-experience-3/the-gutenberg-diagram-in-web-design-e5347c172627

37

Use demarcation lines

A tried-and-true way to reinforce groupings of related information is to outline them or separate them using demarcation lines. Think of these lines as fences that help people recognize boundaries between sections.

When developing instructional material, demarcation lines can provide additional cues to the user to help them navigate the document and organize information. For example, they can:

- Clearly delineate between different content types (e.g., warnings, product care, procedural steps, manufacturer's information)

- Visually group all sub-steps and illustrations associated with a step

- Reinforce the content hierarchy

- Highlight key information in call-outs or sidebars

Demarcation lines can look especially simple and appealing when their color matches the primary color of the document's header (also called a banner). For example, a header's background and associated demarcation lines may be colored blue, making them noticeable without distracting from critical content printed in black or an alert color (e.g., red). Depending on the header's lightness or darkness, the included text, such as in a title, can be light or dark to ensure sufficiently high contrast, which improves legibility.

5. Remove needle

Do not re-use needles. Always remove used needles before using the injector again.

To remove a used needle:
1. Push the needle ejector button (see Figure E).
2. Lift off the used needle.

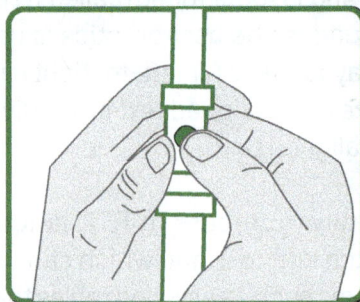

Figure E

The green demarcation lines group the contents of this panel, while the black demarcation line distinguishes the illustration from the accompanying content.

Demarcated sections can also include further subsections. In the previous example, the outlined section contains an outlined illustration, which helps it stand out from the adjacent text. Using outlines with different appearances can represent their relative importance. In this example, the larger section is demarcated using a colored and relatively thick (1 mm) outline, and the subsection uses a thinner (0.25 mm) black outline.

Full outlines (i.e., boxes) look nice when they have rounded corners. The example on the previous page has 5% corner radii.

It is also important to distinguish between sections by including what graphic designers call a gutter — a margin between outlines. In the example on the previous page, the gutter is about 4 mm (1/8 inch). An alternative approach that saves a bit of space is to place content into cells within a grid, whereby adjacent cells share a demarcation line.

38

Continue on back

If your document has two sides, you face two main hurdles in getting users to properly review your instructions. The first is to make sure users realize they should start reading the document on its front side.

The solution to this is obvious. Conspicuously label the sides as *front* and *back* or *side 1* and *side 2*, even if this seems unnecessary in view of other cues about the document's beginning and end.

Such cues might include:

Beginning (Front/Side 1)

- Main document title
- Image of the associated device
- Introductory header and paragraph
- Beginning of a series of numbered steps

End (Back/Side 2)

- Image of the associated device
- Continuation of a series of numbered steps
- Ancillary content that might not usually warrant the user's attention (e.g., manufacturer contact information)

Continues on back ↪

The beginning and end navigation cues listed above may be complemented by such text as "Getting started" versus "Finishing up" and their equivalent.

Seriously, getting users to turn the page over—the second hurdle—can be a big challenge. This is particularly important when the procedural steps continue from side 1 to side 2, creating the potential for readers to mistakenly think they've completed the procedure after following only some steps. There are also cases where all procedural guidance fits on side 1, while side 2 contains more key information, such as how to properly care for the associated product.

Common techniques to get readers to turn the page over include one or more of the following:

- Text stating "continues on back" or the equivalent

- Icon indicating to flip the page

- Graphic suggesting content continues on the other side

Make sure such signs and page turning instructions are conspicuous, perhaps by making them large and colorful.

39
Make instructions visually appealing

You've likely heard the adage "form follows function," which asserts that design should be driven by its functional purpose rather than decoration and ornamentation. In the context of medical product instructions, we believe that function can even create appealing form — that good usability is the driving force behind aesthetically appealing instructions.

The Guggenheim Museum, designed by Frank Lloyd Wright, is an example of form following function while resulting in an arguably gorgeous building. The spiraling levels enable visitors to walk, uninterrupted by staircases, between all levels (Original photo by Kristoffer Arvidsson).

Adding visual appeal is different from adding decoration. Visually appealing instructions are simply pleasing to the eye and can be the outcome of following the kinds of design principles covered in this book and others on typography and document design.

Making instructions visually appealing really is no more difficult than making them unappealing. But you do need to exercise a measure of design sensitivity. The same thing could be said about presentation slides, which can look pleasing or irritating depending on how much effort their composer put into them.

Here are a few design guidelines, some of which are explored in greater detail elsewhere in this book, followed by exemplars.

Maintain consistency. Readers tend to be distracted by misalignments and unintentional differences within a document. Aim to create consistent spacing, sizing, and presentation of visual elements to give the document a professional appearance.

Limit information density. Based on our experience, we advise leaving about 20%-30% blank space on a page for most instructions. Many people have proposed ways of measuring information density on a page, and there is no clear best way to do it. One simple way is to squint when viewing a page and roughly estimate how much space is blank. Using this approach, blocks of text count as content. You do not subtract space between words of the area inside all of the letters.

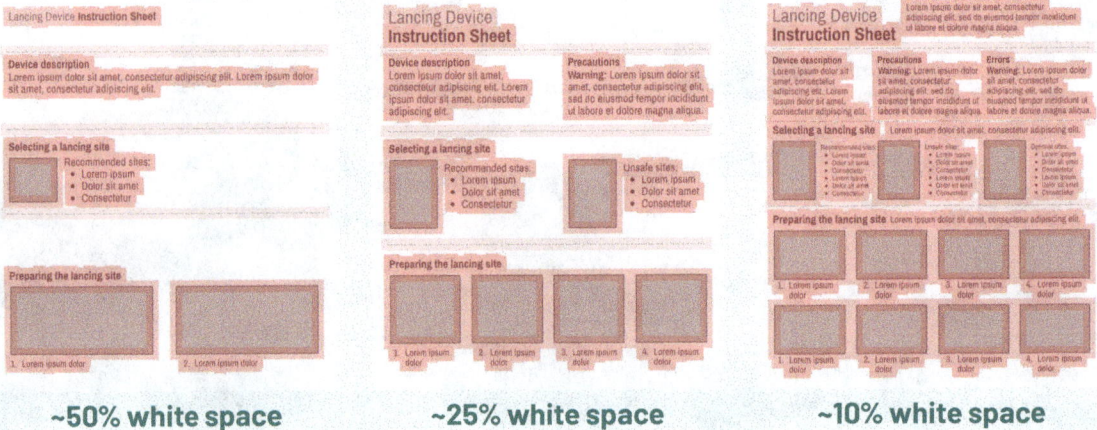

~50% white space **~25% white space** **~10% white space**

Examples of different information densities with red overlays representing the content in the instructions.

Use a grid. Aligning content to a grid can help create a structured, visually consistent arrangement that most users find appealing. That said, avoid using a grid too strictly; having certain content intentionally misaligned with the grid can help it stand out.

Use consistent margins and gutters. Maintaining a consistent space at the document's perimeter (i.e., margin) and between blocks of content (i.e., gutters) gives the document a balanced and more harmonious appearance.

Size headings appropriately. Use headings and sub-headings that vary in size by at least 25% so that users can tell the difference and appreciate the intended information hierarchy. Aim to use proportional scaling based on a consistent ratio, which will ensure that the different headings seem harmonious with each other.

Title	(16 pt)	Title	(28 pt)
Headings	(14 pt)	Headings	(20 pt)
Sub-headings	(13 pt)	Sub-headings	(16 pt)
Body Text	(12 pt)	Body Text	(12 pt)
Captions/Notes	(11 pt)	Captions/Notes	(10 pt)

Two alternative font sizing hierarchies: the example on the left includes barely discernible size differences, while the one on the right includes more distinct sizes for each heading level.

Use helpful graphics. Integrate harmonious graphics where they serve a definitive purpose rather than just decorate. There's almost always an opportunity to communicate graphically in a helpful way, and this has the benefit of "taking the curse off" of an otherwise all-text document. People tend to find all-text instructions unappealing.

Use color effectively. Use color in a limited and meaningful way, as suggested in "Use color effectively" on page 100.

Reinforce branding. Reinforce the manufacturer's brand in an understated manner that adds to visual appeal without reducing legibility.

Demarcate content. Use background fills and gradients instead of always putting content in boxes. Alternatively, use blank space to separate functionally related content instead of always adding a demarcation line.

Limit typography. Using only one or two typefaces will create a more consistent, professional appearance. Using more fonts or a unique font for only one piece of content leads users to associate special meaning to the content that might not be intended. Also, it can subtly add to mental workload when sorting through instructions.

Establish a clear hierarchy. In addition to using headings, arrange, indent, and space content in ways that communicate the relationships between different pieces of content.

Use contrast effectively. High contrast helps ensure legibility. But, selective use of different contrasts can help guide the reader's eye through a document. For example, a medium contrast color pairing can be used for secondary information, while high contrast pairings are reserved for headings and warnings.

Avoid visual clutter. Minimize the use of decorative elements that will make the document appear busier without providing any functional benefits.

Group related content. Where appropriate, "chunk" related blocks of information into fewer groups, thereby simplifying the document's appearance without reducing the amount of content.

Keep in mind that usability testing can do more than confirm your instructions' functional effectiveness. You can also collect feedback on their visual appeal. During an early formative usability test, you can present various concepts (e.g., sample pages) and let the users guide you toward the preferred design.

When you produce visually appealing instructions, you increase the chance that people will pay attention to them, and that's a great benefit. We think you also increase the chance that people will commit more time to reading them.

SIDEBAR
Gestalt principles

Many of the aesthetic guidelines described in this section are based on the so-called Gestalt principles identified by German psychologists in the early 20th century (Gestalt is German for "shape" or "form"). These psychologists sought to understand how humans perceive and make sense of visual stimuli and identified principles that describe how our brains naturally organize and interpret visual information, emphasizing the idea that the whole is greater than the sum of its parts.

In instructional design, applying Gestalt principles—such as proximity, similarity, continuity, closure, and figure-ground—enables designers to create more coherent and visually appealing designs.

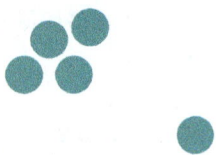
Proximity

Similarity

Continuity

Figure-ground

40

Help users find content of interest

Of course you want users to be able to "navigate" instructions effectively to find information of interest. But users have free will in terms of where they direct their attention.

This is evident from eye tracking studies that examine where people pay attention to content, such as food on grocery store shelves or website articles. In each scenario, grocery store managers and web designers use clever mechanisms that "steer" users to certain content. On a store shelf, it might be placing products level with the consumer's line-of-sight. On a webpage, it might be using visual highlighting and page placement. And for instructions, a savvy designer might highlight certain information that must not be missed.

Eyetracking research[*] suggests that different types of information layouts lead viewers to follow different attention patterns:

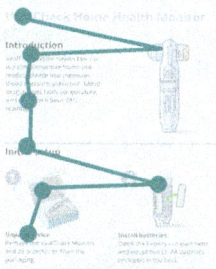

F-Pattern, *where users begin reading on the left side and scan part of the content to the right, possibly continuing down the left side of the page.*

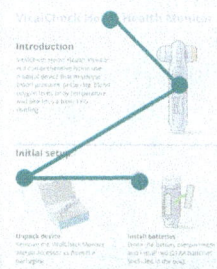

Spotted Pattern, *where users focus on the most visually prominent areas, like bold headings, and read the related content.*

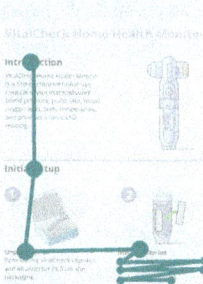

Layer Cake Pattern, *where users focus on conspicuous rows of content, possibly skipping intermediary content.*

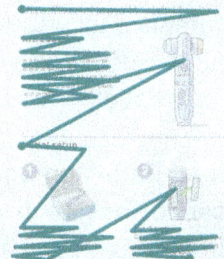

Commitment Pattern, *where users read content thoroughly, much like reading a book. You're likely using this pattern right now!*

---

[*] Kara Pernice. Text Scanning Patterns: Eyetracking Evidence. August 25, 2019.
https://www.nngroup.com/articles/text-scanning-patterns-eyetracking/

Depending on the user and task, users might be predisposed to follow one or more attention patterns. In certain cases, and particularly for instructions presented on a single page, you typically want users to read the content in the intended sequence (e.g., left-to-right and top-down in most languages) and completely. It's a problem if a user skips a key step in a procedure, such as "Dispose of contaminated needle in a sharps container." Such use errors can be linked to document design failures, may interfere with medical product validation (see "Verify and validate the instructions" on page 73), and could lead to harm (e.g., transmission of a blood-borne disease to another person due to needlestick injury).

Therefore, you should avoid inducements for users to jump around a page, reading in what researchers call a "Spotted Pattern," and to facilitate more of a "Commitment Pattern," where all content is read. Some ways to encourage or discourage certain patterns include the use of highlighting and graphics. Many of the other tips included in this book can also help encourage users to read all relevant content (see "Facilitate skimming and scanning" on page 133) and using clear, concise procedural steps (see "Take it step-by-step" on page 90).

In longer, more complex instructions, helping people find information in another location, be it another screen or page of a paper document, can be more challenging than guiding users' attention. For example, "Continue on back" on page 115 addresses the seemingly simple goal of getting people to look at Side 2 of a two-sided document. Therefore, you have to give users convenient ways to determine where to go. Here are some solutions that can help, primarily with lengthier documents:

Index. Indexes are a conventional and effective way to help users find specific information of interest. The key is to include the right keywords — ones that accurately summarize the document's contents and are likely to cross the user's mind when they search for specific information.

Table of contents (TOC). TOCs are a good way to provide an overview of the document's complete contents and in turn help users find information of interest. However, a TOC might only lead a user to find a section or chapter (e.g., Care and Maintenance) as opposed to the exact content of interest (e.g., Replacing batteries).

Page numbers. Page numbers are a basic navigation aid and it's obvious to include them. Take care to include them in a consistent location on every page and make them large enough that users can find them quickly. Also, make sure their print contrasts sufficiently with the background. Sometimes, this might require changing the color of the page number (e.g., using a black number on pages with a white background and a white number on pages with a dark blue background).

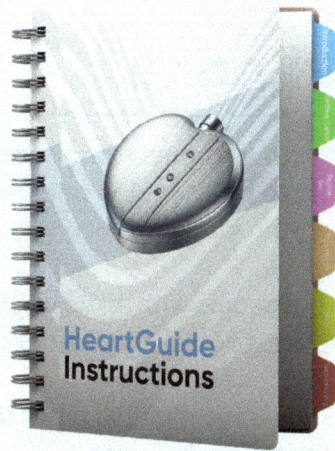

Example of a tabbed instructional document that provides quick access to each section of the document.

Tabs. In physical instructions, tabs provide a visual overview of the document's sections, as well as an efficient way to access specific sections without a lot of page flipping. Tabs can be especially helpful when they're color-coded (i.e., with each section having a uniquely colored tab) and when the tabs are printed on thicker card stock or laminated to ensure that they withstand extensive use. Tab-like buttons in digital instructions can provide similar benefits as well.

Cross-references and hyperlinks. It's best to provide information at the location where users will need it. But sometimes, especially with more complex products that require longer instructions, related information might be distributed across multiple sections to minimize redundancy. In such cases, cross-references can guide users to find and read information that might complement the content they're currently reading. In digital instructions, hyperlinks perform the same purpose.

Search. When using digital instructions, the most efficient way to find information of interest is likely a search field, in which users can type one or more words related to their topic of interest. Basic search functionality, such as that used in PDF viewers, will enable users to find instances of those exact words in the document, while more sophisticated search features, such as those found in custom learning systems, can find relevant content using synonyms to the entered search term.

Bear in mind that different users have different preferences for finding content of interest. The best way to help the widest range of users is usually to employ some combination of the options described above.

41

Sizing and styling text

Imagine an instruction sheet with extremely small text. For people with normal or fully corrected vision, it's going to be annoying or even impossible to read without a reading aid, such as a magnifying glass.

The reading challenge worsens for people with some visual impairment (e.g., glaucoma) or who are farsighted (i.e., have presbyopia) and are not wearing their reading glasses.

Below, we present the same text passage multiple times to give you a sense for how text becomes illegible at a certain size for each person.

Sample text (using the Lato typeface, which has an x-height to font size ratio of .59)	Font size	Visual angle from 40 cm distance
Do not use the prefilled syringe if the medicine is cloudy or discolored or contains flakes.	18 pt	32 arc-minutes
Do not use the prefilled syringe if the medicine is cloudy or discolored or contains flakes.	14 pt	25 arc-minutes
Do not use the prefilled syringe if the medicine is cloudy or discolored or contains flakes.	12 pt	21 arc-minutes
Do not use the prefilled syringe if the medicine is cloudy or discolored or contains flakes.	10 pt	18 arc-minutes
Do not use the prefilled syringe if the medicine is cloudy or discolored or contains flakes.	8 pt	14 arc-minutes
Do not use the prefilled syringe if the medicine is cloudy or discolored or contains flakes.	6 pt	10 arc-minutes

Obviously, using larger text would be the smarter way to go, but this would require more space on the page, resulting in larger or longer instructions. There is a tradeoff. Still, you certainly want to ensure that the instructions' content can at least be read, even if the reader thinks it could be larger still.

So, what text sizes ensure legibility? We provide a detailed formula for calculating minimum text sizes based on reading distance at the end of this section. But, in most cases you can rely on the following rules of thumb:

Document format	Approximate viewing distance	Minimum text size
Digital instructions viewed on a smartphone / small printed instructions (e.g., postcard-sized quick reference card)	30 cm (12 inches)	10 point
Medium-sized printed instructions (e.g., a single sheet, pamphlet, or booklet)	40 cm (16 inches)	12 point
Digital instructions viewed on a laptop or desktop display	70 cm (28 inches)	16 point
Large printed instructions (e.g., poster or oversized pamphlet)	100 cm (40 inches)	24 point

Determining a minimum font size for your document is a good start. But you should also consider the following variables to optimize your document's legibility:

Typefaces. There are hundreds of typefaces designed to meet various functional and aesthetic goals. The safe action is usually to use a sans serif font, such as Helvetica, Arial, Calibri, and others. These typefaces look simple and forsake any flourishes that add personality but might interfere with legibility when viewed on a lower resolution digital screen. That said, serif typefaces such as Times New Roman, also offer good legibility and can help guide the eye from word to word, especially in printed documents.

Fonts. A stylistic variant of a particular typeface is known as a font. For example, the popular Helvetica typeface is available as multiple fonts that feature different character thicknesses (also known as weights) or slopes (also known as italic or oblique styles). Using bold fonts can draw attention to text. Similarly, using italic fonts can help distinguish one type of word from another (e.g., when defining a product component). Other fonts, such as condensed, thin, or wide variants can add stylistic diversity, but often hinder legibility.

Alignment. Text can be categorized as being left-aligned, right-aligned, centered, or justified. For languages that are written from left to right, left-aligned text is often the most legible because it provides a consistent starting point for the eye. Conversely, right-aligned and centered text can make it more difficult for the eye to locate the start of the next line, thus reducing legibility. Justified text, where both the start and end of the text block are aligned, might reduce legibility because it can create uneven spaces between words that can make text more challenging to read.

Left-aligned	*Center-aligned*	*Right-aligned*
Ensure that the cuff is secured around the upper arm, with the sensor placed over the brachial artery.	Ensure that the cuff is secured around the upper arm, with the sensor placed over the brachial artery.	Ensure that the cuff is secured around the upper arm, with the sensor placed over the brachial artery.

Letter spacing. The distance between individual letters is known as kerning. Insufficient spacing between letters can make them appear blended together. For example, when the letters "c" and "l" are placed too closely they are likely to be misread as "d", which could cause the word "clot" to be misread as "dot." Because readers often recognize words as whole shapes, rather than just individual letters, excess kerning makes it harder to recognize words quickly and accurately. It's best to keep the default kerning, which is usually a setting of 0. That said, when using a relatively small font, slightly increasing the kerning can reduce potential for letters to appear blended together.

Tight kerning *(-200 ems)*	*Default kerning* *(0 ems)*	*Loose kerning* *(250 ems)*
bloodclot	blood clot	b l o o d c l o t

Word spacing. You might expect the space between words to be of a standardized length. But, the so-called "word space" — the empty character that appears when you press the spacebar on your keyboard — actually varies in length depending on the typeface. Furthermore, designers can adjust the word space when setting text on a page. Spacing words too closely or too far apart can slow reading speed and increase potential for reading errors. A good rule of thumb is to make the space between words approximately the width of a lowercase 'n' in the typeface that's used.

Line spacing. The vertical space between two lines of text is known as leading (it rhymes with sledding). If lines of text are too far apart, some readers may find it difficult to move smoothly from one line to the next. In general, line spacing in the range of 120%-150% of the text's point size should make text legible and pleasant to read.

Tight leading *(60%)*	*Well-balanced leading* *(150%)*	*Loose leading* *(200%)*
Ensure that the cuff is secured around the upper arm, with the sensor placed over the brachial artery.	Ensure that the cuff is secured around the upper arm, with the sensor placed over the brachial artery.	Ensure that the cuff is secured around the upper arm, with the sensor placed over the brachial artery.

Text hierarchy. You'll likely want to use different size text to indicate information hierarchy, such as differentiating headings from body text. The rule of thumb is to start with the body text and then increase character sizes from there by at least 25% per size increase. Therefore, if you are starting with 12 point text, the next two sizes could be 16 and 20. Even larger proportional jumps can work well. The key is to make sure people notice the difference and associate meaning to it. A difference less than 25% might not be noticeable to the reader's eye. This relates to the just noticeable difference principle, which is that people can overlook a small change; people cannot discern the difference and so it simply does not capture their attention.

Calculating a minimum font size

ANSI/AAMI HE75:2009 (R2018) *Human factors engineering - Design of medical devices* recommends determining text size based on the expected viewing distance.

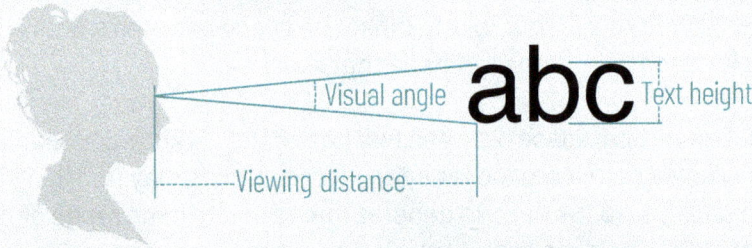

HE75 suggests using a minimum text size that subtends a visual angle of at least 16 arc-minutes. We can use the following formula to calculate a text height (in inches) based on the visual angle (in arc-minutes) and viewing distance (in inches):

$$\text{Text height} = \tan\left(\frac{\text{Visual Angle} \times \pi}{180 \times 60}\right) \times \text{Distance}$$

Let's assume a viewing distance of 16 inches (a typical reading distance when holding a document at arm's length) and a visual angle of 22 arc-minutes (an appropriate target for comfortable reading). The formula produces a text height of 0.102 inches, which is equivalent to a 7 point font (one inch equals 72 points).

You'll be correct to note that this is quite a small size. That's because this size actually refers to the typeface's "x-height," which is the height of a lowercase letter that doesn't have any ascenders or descenders. However, when you set a font's size in a word processor or design program, you're actually setting its body height, which is the distance from the typeface's topmost position (ascender) to its bottom-most position (descender).

To determine the body height corresponding to our calculated x-height, we need to know the ratio of the typeface's x-height to its body height. Let's assume our typeface is Helvetica, which has an x-height ratio of 0.54: We divide our calculated x-height of 7 points and divide it by 0.54, resulting in a body height (i.e., font size) of 13 points.

42
Referencing figures and tables

Figures and text go together like peanut butter and jelly. They're both nice on their own, but work even better when you put them together.

A table or figure, which could be an illustration, chart, photograph, or other visual representation of content, can help clarify a portion of the text or even eliminate the need to include text to describe something that can be depicted more simply. Conversely, text can help clarify details that might not be visible in a figure.

But, in order to work well together, figures and text must be presented in a way that makes their relationship to each other clear.

The best practice is to present one figure next to its matching text (e.g., sentence or paragraph). For example, the following illustration of a cap being removed from a syringe placed next to its associated prompt.

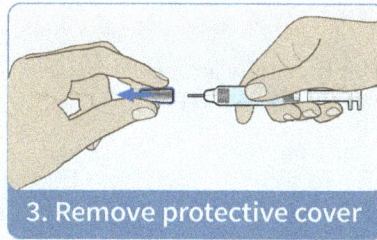

3. Remove protective cover

But, such layouts aren't always possible due to limited space or large amounts of text. The next best option is to include references. Consider the following frame that contains multiple sentences and one illustration.

Using the syringe

1. Wash your hands thoroughly with soap and water.
2. Gather all necessary materials, including the syringe, medication vial, and alcohol swabs.
3. Remove the protective cover (**Figure A**)
4. Draw the desired amount of medication into the syringe by pulling back the plunger.
5. Tap the syringe gently to remove any air bubbles, and expel them by pushing the plunger slightly.
6. Administer the injection as directed, then safely dispose of the syringe in a sharps container.

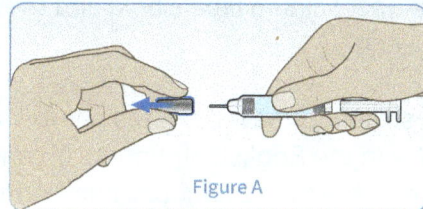

Figure A

Is it necessary to refer to the illustration in the text by name (e.g., "See Figure A") and then add an identifying label (e.g., "Figure A") next to the illustration?

You might question the need to point to a figure when its relationship to the text seems obvious. But, the relationship might not be obvious to the reader. In the second example on the previous page, the figure reinforces the instruction to "keep the needle straight as you screw it on." It is possible, if not likely, that the reader might incorrectly associate the figure with another portion of the text in the same frame.

Below, the frame includes several instructions and two figures. Again, the references might be unnecessary because of the figures' adjacency to the associated text. Nonetheless, the references help avoid confusion.

Delivering a dose

1. Open white cap. **(Figure A)**

2. Close your lips around mouthpiece, keeping your tongue below it.

3. Breathe in deeply as fully as you comfortably can. **(Figure B)**

4. Remove inhaler from mouth, hold breath for 10 seconds, then breathe out.

5. Close white cap to prepare inhaler for your next puff.

Figure A

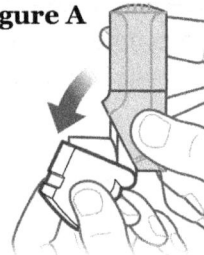

Figure B

The example shown above also demonstrates the value of making figure references and their associated labels bold and the same font size. This creates a helpful, visual equivalence.

If the instructions have a limited number of figures or tables (≤ 26, which is the number of letters in the English alphabet), go ahead and label them A, B, and so forth. When there are more than 26 figures, numbering is the way to go (i.e., 1, 2, and so forth). With the limited number of figures and tables that could appear in most instructions, you can probably dismiss more intricate numbering, such as Table 1.1, Table 1.2 through Table 7.1, and Table 7.2, for example.

43

Sidebars

When people converse, they might stay on point or go off on a tangent. "Oh, by the way..." is a verbal turn signal indicating that the conversation is shifting focus. Sidebars in documents are a similar shift in focus. They provide information that is useful to know but perhaps not essential.

Do sidebars belong in instructions? Perhaps not, because the instructions should focus on information essential to the safe and effective use of a medical device. But perhaps so, because in some cases secondary information can provide additional context that might enhance a user's knowledge or use of the device.

Here are a few examples of the kind of information that might be appropriate for a sidebar:

- More detailed explanations of the reason for the need to perform a particular step. For example, explaining that holding a needle in the skin for 10 seconds after a drug infusion decreases the likelihood of drug leakage from the injection site.

- Suggestions for performing certain steps more easily or efficiently. For example, encouraging users to prime multiple infusion sets simultaneously when setting up multiple infusion pumps.

- Notification that the device is complemented by a helpful accessory or mobile application. For example, providing the QR code to download an application that can track the frequency and duration of device use.

- References to resources that might provide additional information to the users. For example, pointing the user to instructional material for auxiliary components that inter-operate with the primary device being used.

Don't assume that all users will follow your sidebar. Users should be able to disregard the sidebar's contents without it negatively affecting their ability to follow the primary instructions.

Sidebars should be visually subordinate to the instructions' primary content. You can do this by:

- Presenting sidebars in smaller "blocks," perhaps with a width that is 1/4 to 1/3 of the primary content.

- Giving sidebars a slightly subdued appearance, perhaps with a light gray or blue background but making sure they still contain good figure-to-ground contrast. For more detail on effective color choice, see "Use color effectively" on page 100.

- Adding an outline and indenting the sidebar from the primary content.

- Including a header that reinforces the secondary nature of the sidebar's contents (e.g., "Additional information" or "Helpful tips.")

SIDEBAR
A historical sidebar

The term sidebar originates in 18th century England, where it referred to "side-bar rules" — informal legal rulings made by attorneys sitting at a bar on the side of the court in Westminster Hall (the chief court of England at the time).*

The term is still used in contemporary legal practice to describe in-court conversations between the judge and attorneys, but out of the jury's earshot.†

In the early 20th century, newspaper designers adopted the term to describe relatively short, secondary articles that appear alongside the primary articles on a newspaper page.‡

An etching from 1733 depicting proceedings at Westminster Hall, with a sidebar conversation shown in the bottom right corner.

---

*   https://www.merriam-webster.com/dictionary/side-bar%20rule
†   https://www.merriam-webster.com/dictionary/sidebar
‡   https://www.etymonline.com/word/sidebar#:~:text=sidebar%20(n.),in%20mechanics%2C%20law%2C%20etc

44

Facilitate skimming and scanning

Have you been carefully reading, skimming, or scanning this book? While of course we hope you've mulled over every word we've written ad nauseam, realistically, you've done a mix of the three.

Carefully reading means working your way from the top of the page to the bottom and not skipping anything. When skimming, you read sentence by sentence but perhaps skip some words that you know or assume are not crucial to getting the main point. Scanning involves jumping around the page, perhaps looking for something specific or getting a general but not detailed sense of the page's content.

Instruction designers cannot, and should not, expect to control reader behavior. A reader will do as they wish or in accordance with other requirements, such as carefully reading content during a reading comprehension test that they would otherwise skim.

For example, consider a person engaged in an online training delivered via a learning management app. If the app allows readers to confirm having read the instructional content by pressing a "Confirm" button, then readers are more likely to quickly scan the content. But, if the system requires readers to pass a comprehension quiz before receiving credit for the training, readers will likely read the content more carefully.

That said, instruction designers can help skimmers and scanners get the most out of their limited attention to a document and facilitate safe and effective use of medical devices. Here are some tips for making instructions more skimmable and scannable:

- Keep sentences short
- Keep paragraphs short (perhaps no more than 3 sentences)
- Remove unnecessary words
- Include meaningful headings and sub-headings
- Highlight key information, such as a warning, by increasing blank space around the content or differentiating it with demarcation lines
- Include simple, captioned illustrations to complement or replace text
- Use bulleted lists and tables
- Present content in a logical order
- Present key information in the first sentence of a paragraph
- Use relatively large fonts, at minimum for the key content
- Consider presenting a summary of content ahead of detailed content
- Number things when relevant, using numerals (e.g., 8) rather than spelled-out numbers (e.g., eight)

CHAPTER FOUR
WRITING CONSIDERATIONS

45

Use good grammar when directing users

"To be or not to be" is a difficult question and a great way to start a Shakespearean soliloquy. It's also a good example of a sentence starting with an infinitive verb — the unconjugated form of the verb, usually preceded with the word "to."

With all due respect to Shakespeare, for the purposes of this section, the more relevant question is: "To use the infinitive, or not to use the infinitive?"

Let's consider a few sentences that start with an infinitive phrase:

- To set a dose, rotate the dial to the desired number (e.g., 10 to deliver 10 units of insulin)

- To turn off the monitor, press the Power button for 3 seconds

- To save the new settings, press the Enter key

The sentences begin by stating the end results and then describe the action to achieve it. The advantage is that users can read what will happen before they take the action. This is useful when you want the user to proceed carefully and be aware of the results and potential consequences of their action. It's also useful when you're describing alternative actions that the user can take at a given time. It avoids the scenario in which a user takes the action right away according to the instruction but without paying attention to the rest of the sentence. As a result, this approach can help prevent errors of commission, whereby the user took deliberate action but expected a different outcome.

The disadvantage of starting sentences with an infinitive is that it reverses the natural order of events, presenting the result before the action needed to achieve it. As a consequence, readers must exert more effort to understand the proper sequence of events before executing a task. This could lead to delays or confusion in high-stress scenarios and negatively impact readability for people with low literacy, cognitive impairments, or those in highly distracting use environments. Moreover, because the use of the infinitive "buries the lead" of the sentence, users might skim or skip sentences they consider less important, potentially missing crucial steps.

Air traffic controllers often speak using imperative phrases, such as "Turn right on Bravo, Link 21 join Alpha, hold at MORRA."

When an action is required, as opposed to optional, it's better to begin the instruction with the imperative, rather than infinitive, form of the verb. Consider the following revisions to the sample sentences presented previously:

- Rotate the dial to the desired number (e.g., 10 to deliver 10 units of insulin) to set the dose

- Press the Power button for 3 seconds to turn off the monitor

- Press the Enter key to save the new setting

All of these examples have one thing in common: They lack an explicit subject. As suggested in the table on the next page, excluding a subject from the sentence is acceptable because the instructions are speaking directly to the user. Accordingly, the user understands that they are supposed to follow the instruction.

Maintaining a consistent voice

In many forms of writing, using diverse verb forms is an effective way to add variety and pique user interest. For instructions, mixing verb forms haphazardly produces unclear and confusing texts. To maintain consistency in your instructions, consider creating a style guide that defines which voice to use for particular types of actions, such as the following table:

Action type	Suggested voice	Example instruction
Optional action	Infinitive	To view additional settings, click on the menu icon.
Alternative actions	Infinitive	To cancel the infusion, tap "Cancel." To delay the infusion, tap "Delay."
Urgent action	Imperative	Press "Start" to begin the infusion.
Multi-step actions	Imperative	Press and hold the "Prime" button until the reservoir is filled with fluid.

46
Use plain language

Reading instructions shouldn't feel like taking a vocabulary test. It's not that advanced vocabulary words are verboten. But, when editing text, you should ask if simple, plain wording would be better for most readers. (Perhaps "forbidden" would have been a better choice than the more esoteric "verboten?")

In general, we think simpler words are far better, even for sophisticated readers, unless using a simple word in place of a more advanced word obfuscates, or rather, confuses.

Different tools are available to measure how plainly content is worded. For example, the ubiquitous Microsoft Word has a built-in tool based on the Flesch-Kincaid Grade Level formula for estimating a document's reading grade level. While such tools can help you estimate a given document's reading level, they won't help you write simple text in the first place. For that you need a "keep it simple" mindset when writing, and then be a fierce editor: simplifying complex sentences, using simpler vocabulary, and eliminating jargon.

Three things to remember:

1 Simplify

2 Simplify

3 Simplify

On the next page, let's look at two alternative versions of the same instructions for measuring blood pressure using an automatic blood pressure measurement device. An original version at a college reading level, and a revised version, rewritten for a fifth grade reading level.

College reading level

Flesch-Kincaid Grade Level: 14.7
Flesch Reading Ease Score: 30.42

1. Prior to initiating blood pressure measurement, abstain from consuming food and engaging in smoking or strenuous activities; additionally, ensure your bladder is voided.

2. Find a quiet, comfortable location and allow yourself to sit, back supported and feet evenly rested on the floor, for a duration of five minutes to achieve a state of relaxation.

3. Position the monitoring cuff on the exposed surface of your upper arm, precisely one inch above the elbow's curvature.

4. It is vital to ensure the cuff is secure yet not overly tight — a gap sufficient to accommodate two of your fingers should be present.

5. Rest your arm on a surface that is flat and stable, ensuring the upper arm is aligned with your heart to ensure accurate reading.

6. Depress the "Start" button on your device, simultaneously maintaining an absolute state of stillness and refraining from conversation.

7. Once the cuff has completed its inflation-deflation cycle, accurately document the systolic (first/higher number) and diastolic (second/lower number) pressures as shown on the device monitor.

Fifth-grade reading level

Flesch-Kincaid Grade Level: [*] *2.62*
Flesch Reading Ease Score: 94.85

1. Don't eat or smoke before checking your blood pressure. Make sure you have used the bathroom.

2. Sit and rest for 5 minutes. Your back should lean on something, and your feet should touch the ground.

3. Put the blood pressure cuff on your upper arm, above your elbow.

4. The cuff should be snug but not too tight. You should be able to put two fingers under it.

5. Put your arm on something flat, like a table. Your arm should be as high as your heart.

6. Push the "Start" button on your machine and don't move or talk.

7. Write down the numbers that show up on the machine after the cuff gets small again.

---

[*]   Flesch Kincaid Calculator. Good Calculators. https://goodcalculators.com/flesch-kincaid-calculator/

Note that the fifth grade wording is more concise and uses plainer words than the college level text. This highlights two strategies you can use to simplify your writing:

- Review your text and identify words to eliminate (see "Write short sentences" on page 149)

- Substitute words with simpler alternatives

The old-fashioned way to find simpler words is to use a thesaurus and select synonyms that are more likely to be familiar to your target audience. Alternatively, you can use AI tools like ChatGPT and Claude to suggest simpler alternatives (for example, by asking "What would be an alternative plain language term for [complex word] that is likely to be understood by individuals with low literacy?").

When substituting simpler words, make sure the simpler word does not affect users' interpretation of the instructional text. For example, the word "disinfectant" might be familiar to users than the word "antiseptic" but could be misinterpreted as referring to a general cleaning agent rather than one intended only for cleaning wounds.

SIDEBAR
SMOG can clear up medical writing

In addition to the Flesch-Kincaid Grade Level formula, other reading level formulas include the Gunning Fog Index, Dale-Chall Readability Score, and the Simple Measure of Gobbledygook (SMOG). All of these formulas are based on factors like sentence length, word length, and the complexity of vocabulary used, but each formula weighs these factors in a different way.

The SMOG formula, in particular, has been shown to be especially effective for health-related documents. A study focusing on health-related documents* found SMOG to be the most reliable among the formulas evaluated. SMOG places significant emphasis on polysyllabic or complex words (those with three or more syllables), which are often abundant in health literature. By focusing on these complex words, SMOG can provide a more accurate estimate of how understandable health-related documents are for a typical reader. This makes it a particularly useful tool for writers of medical product instructions.

You can find several SMOG calculators online, such as the one available at this URL: https://charactercalculator.com/smog-readability/

---

* Wang, Lih-Wern & Miller, Michael & Schmitt, Michael & Wen, Frances. (2013). Assessing Readability Formula Differences with Written Health Information Materials: Application, Results, and Recommendations. Research in social & administrative pharmacy : RSAP. 9. 503-516. 10.1016/j.sapharm.2012.05.009.

47
Avoid ambiguity

Through poet's quill and lawyer's ink,
ambiguity unites,
crafting subtle shades of meaning,
in creative and legal rites.

But when it comes to tools that heal,
precision is our creed,
for vague instructions cause mistakes,
when users they mislead.

You probably weren't expecting to explicate any poetry in this book, so we'll give you a hand: the poetic stanza above is our joking way of saying that ambiguity has its place in many forms of writing. Poets and lawyers, for example, may both use ambiguity to allow for subtle and divergent interpretations of their writing.

What's not a joke is reading instructions that are ambiguous. They can confuse people. They can slow down urgent tasks. They can trigger mistakes.

Let's state it bluntly: Ambiguous instructions are bad.

Most often, ambiguity is caused by word choices and sentence structures that users can misinterpret. Consider this example:

Attach the dialyzer tubes.

In this case, it's unclear — at least to a first-time user — to what they should attach the dialyzer tubes and from where the tubes are originating. You can resolve this ambiguity by adding a few words:

Connect the dialysate tubes to the dialyzer.

Still, there remains a question: to where on the dialyzer are the tubes being connected? You can reduce ambiguity with more specificity:

Connect the dialysate tubes to the dialyzer's inlet and outlet ports.

Much better, but which tube is connected to which inlet and outlet port? We believe in economy of words, but two instructions are probably better than one in this case:

Connect the blue, used dialysate tube to the dialyzer's upper outlet port. Connect the red, fresh dialysate tube to the dialyzer's lower inlet port.

To further reduce ambiguity, the instructions could specify that the inlet and outlet ports are on the dialyzer's side, but an accompanying diagram could make this quite clear as well.

Also, the writer might know that the dialysate tube cannot be connected to the blood input and output ports because they are physically incompatible to be inherently safe. However, omitting the detailed instruction and relying on the device's design to prompt correct use could induce confusion and result in delayed treatment or even a failure to connect the tubes altogether.

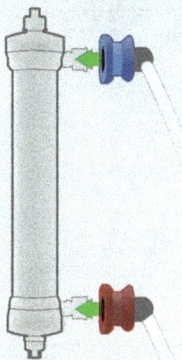

Connect the blue, used dialysate tube to the dialyzer's upper outlet port.

Connect the red, fresh dialysate tube to the dialyzer's lower inlet port.

Another way to prevent ambiguity is to avoid words with multiple meanings (i.e., homonyms). Consider this example:

Place the dialysate bag on the warmer with the right side pointing up.

This advisory is flawed because it uses the term "right," which could mean correct or right-hand facing. It's also ambiguous because some users might misinterpret "right side" as referring to the warmer's right side. Here's an improved version:

Place the dialysate bag on the warmer with the bag's labeled side facing up.

In case you're curious, the bag's label should face up so the user can confirm the contents (fluid name, concentration, amount) and ensure the bag has not expired, tasks which might or might not have been performed during prior handling tasks.

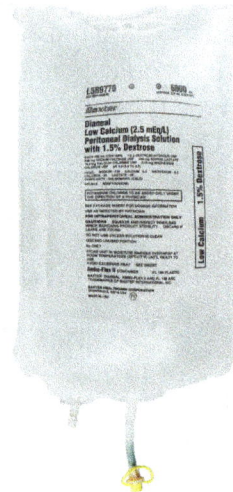

In the photo on the left, a dialysate bag is shown atop the dialysis machine's warmer. The photo on the right shows the dialysate bag and its labeling. (Images courtesy of Vantive Health LLC)

One more way to reduce ambiguity is to avoid misplaced modifiers. The solution is to keep modifiers and the word they modify close together. Consider this example:

People who disinfect their skin rarely are more vulnerable to infection.

As you see, the word *rarely* is problematic. Instead of learning that infrequent skin disinfection increases the chance of infection, a reader could conclude the opposite — that infrequent skin disinfection does not often lead to increased vulnerability. The simple solution is to move the word *rarely*:

People who rarely disinfect their skin are more vulnerable to infection.

48

Edit ~~good.~~ ~~Edit well.~~ Edit

The perfect first draft is a unicorn. Something that you can imagine but not something found in nature. Mozart was purported to be a one-of-a-kind musical genius who wrote music perfectly the first time without correction. False.

Mozart is often presented as a near perfect personification of the inspiration theory of creativity. Nowhere is this more visible than in the movie *Amadeus*.

Amadeus depicts Mozart's bristling relationship with Antonio Salieri, who saw himself as Mozart's rival. In one scene, Salieri gazes at a finished Mozart composition and states: "Astounding! It was actually beyond belief. These were first and only drafts of music, yet they showed no corrections of any kind... Here again was the very voice of God!"

The film's portrayal has its origins in a letter written by Mozart that was published in 1815 by a leading German music magazine. In the letter, Mozart explained his composition process: "Provided I am not disturbed, my subject enlarges itself, becomes methodized and defined, and the whole, though it be long, stands almost finished and complete in my mind, so that I can survey it, like a fine picture or a beautiful statue, at a glance. Nor do I hear in my imagination the parts successively, but I hear them, as it were, all at once."

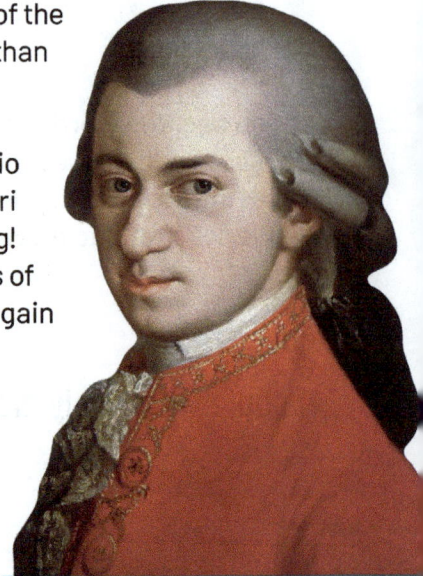

Wolfgang Amadeus Mozart

There is one problem with Mozart's letter: it was a forgery.

In reality, Mozart worked long hours in a highly iterative, back-breaking process. He described a set of string quartets he composed as the "fruit of long and laborious effort."[*]

So, don't beat yourself if your first draft of instructions doesn't flow effortlessly from your quill or keyboard. Good results usually occur through hard work and the kind of iterative process the maestro himself embraced.

As you edit, we suggest keeping these tips in mind:

- Use plain language (see "Use plain language" on page 139).

---

[*] Allen Gannett. The Myth. Inside Design. June 26, 2018. https://www.invisionapp.com/inside-design/mozart-creative-success/

- "Speak" in simple terms. Ask yourself if there is a way to simplify what you already wrote down. For example, replace "In advance of initiating the sensor calibration sequence..." with "Before calibrating the sensor..."

- Imagine you get a dollar (or a fair amount in your local currency) for each word you remove. Or think about eliminating words like trimming hedges. If you do a good job of cutting, your instructions will look (or read) better.

- Write in the active voice (see "Use active voice construction" on page 150).

- Try to limit paragraphs to 2-3 sentences. Reading instructions can be a bit like climbing or descending a staircase. It is comfortable to proceed several steps and then take a brief breather on a level platform (i.e., landing, required by some codes to be every 12 vertical feet*) before going further. The space between paragraphs gives users a breather, so to speak.

- Limit each paragraph to one main point, just as you might have been taught in seventh grade.

- Read the instructions out loud. As you read aloud, you'll likely stumble upon a grammatical problem, missing word, odd phrase, or a typo you overlooked.

- Take a fresh look. What seems like a great piece of writing on Tuesday might seem in need of edits on Friday. So, taking a second editing pass after some time gives you greater objectivity.

- Read a printed copy. Editing a printed document sometimes exposes opportunities for improvement that escaped notice on the screen. It's difficult to explain but true for many writers. (We know, this isn't environmentally friendly, so practice this tip in moderation.)

- Edit at multiple stages of design. As we discuss in "Develop preliminary instructions" on page 61, you can begin writing and editing your instructional text before laying out the instructions or complementing the text with visuals. Edits during this phase can help ensure that the text is clear and understandable on its own. But, it helps to re-edit once the text is incorporated into the draft and final versions of the designed instructions. Edits during these later phases help reveal new opportunities for simplifying the text or harmonizing it with the accompanying visuals.

- Use your word processor's built-in grammar and spelling alerts.

---

* 5311.5.4 Landings for Stairways. https://www.mass.gov/doc/780-cmr-state-board-of-build-ing-regulations-and-standards-building-planning-for-single-and-two/download

Think of editing as hedge trimming.

- Use AI-based tools to review and simplify your text. As we write this, AI-based tools are not yet a substitute for capable human writers. But, they can already serve as a helpful aide. Using large language models, such as ChatGPT, to analyze and simplify instructions can help reveal opportunities for improvement.

- Have other people edit your work. This might seem obvious, but a confident writer who is pressed for time might skip this step at their own peril.

49

Write short sentences

In middle school, if not earlier, you probably were discouraged from writing long sentences because the reader's attention can lapse when sentences get too long, a longer sentence often combines more than one idea, leaving the reader to have to sort things out in a manner that does not always lead to full understanding, and understanding can be compromised when paragraphs get to be too long, such as when they include perhaps 5-10 sentences (and perhaps even more) in a way that seems to go on and on.

We hope you already got the point. Long, run-on sentences are bad. To make the above sample worse, we mixed in some passive voice: "...you probably were discouraged to write long sentences..."(see "Use active voice construction" on page 150). We also ended a sentence with a preposition. Horrors.

So, let's try it again. Your middle school teacher probably discouraged you from writing long sentences. After all, long sentences challenge a reader's attention span. Often, they combine multiple ideas, which can be confusing. Therefore, keep sentences short and keep paragraphs short — perhaps 2-3 sentences.

One way to drive toward short sentences is to cull words from first drafts. Imagine that you get a reward for each word you remove from a sentence. Applying this mindset, you'll be amazed how long sentences easily reduce to shorter, better ones. Here are some examples from hypothetical, medical-device-oriented instructions. You can judge whether it is better to eliminate articles of speech, such as "the," for brevity's sake and when space for words is limited. Yellow highlights indicate words that can be eliminated.

Original 25 words	18 words	16 words
Once you have successfully completed the injection, you should then carefully remove the needle and then be sure to place it in a sharps container.	When you are making a connection, it is very important that you do not kink the tube.	Place the electrode on the left side of the patient's chest, 2 inches above the nipple.

Shortened 10 words	5 words	12 words
Remove the needle and place it in a sharps container.	Do not kink the tube.	Place electrode on left side of patient's chest, 2 inches above nipple.

Be careful to not introduce ambiguity when writing concise instructions. At the end of the day, clarity should be prioritized over brevity. See "Avoid ambiguity" on page 142 for more detail.

50
Use active voice construction

A simple way to increase your instructions' clarity and precision is to use the "active voice" construction, which emphasizes the subject and verb in a sentence.

Consider the following instruction on how to use a metered-dose inhaler:

A deep breath shall be taken by the user to get the medicine into the lungs.

Is there anything grammatically wrong with this instruction? Technically, no. But the instruction is vague. It sounds awkward and indirect. This is because it uses the passive voice; a sentence structure in which the object (the deep breath) has an action (being taken) performed upon it by a subject (the user).

Now consider this alternative:

Take a deep breath to get the medicine into your lungs.

It sounds better, doesn't it? This is because the subject (the user) is performing an action (taking a deep breath). In the above case, the subject is implied rather than stated explicitly. A more explicit but also wordier sentence would be, "Now, you should take a deep breath to get the medicine into your lungs."

In other words: Active voice tends to focus on the user and the action they should take, whereas passive voice tends to focus on the product and the action taken with it by the user.

Here are some more examples of passive voice construction being converted into active voice construction, for the better.

Passive voice	Active voice
Your doctor may be asked about which inhaler spacer chamber will work best with your inhaler.	Ask your doctor which inhaler spacer chamber will work best with your inhaler.
The inhaler spacer chamber should be washed once a week.	You should wash the inhaler spacer chamber once a week.
The number of doses remaining in the inhaler is shown in the counter window.	The counter window shows the number of remaining doses.
The device can be turned on by having its power button pressed.	Press the power button to turn on the device.
The adhesive backing should be removed before the patch is applied.	Remove the adhesive backing before you apply the patch.
The treatment plan shall be discussed.	Discuss the treatment plan with your doctor.

51

Permissive versus prohibitive phrasing

When someone uses a medical device, there are usually a bunch of things they should and shouldn't do. The instruction writer has an obligation to make it clear which is which.

Consider the need for the medical device user to have clean hands when they operate the device. Which of these directions seems better to you?

Option 1
Wash your hands well with soap and water before removing the mouthpiece from its package.

Option 2
Do not remove the mouthpiece from its package until after you have washed your hands with soap and water.

We like Option 1 because it sounds direct and natural. It presents events in the order of occurrence. It has a pleasant, permissive quality to it. However, Option 2 communicates the prohibitive message "Do not touch the mouthpiece" right at the start.

Both options are likely to work in terms of guiding people to act safely. That said, we generally favor telling people exactly what to do rather than telling them what not to do. It's like giving driving directions. You should tell the driver to turn right after passing the red barn rather than telling the driver to not to turn right until they pass the red barn.

However, we think there is a third, better option. That is to avoid compound messages. Notice how both Options 1 and 2 bundle the message to wash one's hands and to do so before touching the mouthpiece. The third option is to separate the two messages, as shown to the right.

If warranted by the use-related risk, you could present a warning in a special format, as illustrated in Option 3.

Option 3

1. Wash your hands with soap and water.

2. Remove the mouthpiece from its package.

WARNING
Placing your mouth on contaminated mouthpiece can cause illness.

CHAPTER FIVE
ILLUSTRATION CONSIDERATIONS

52

Use intuitive graphics

The value of making graphics such as illustrations and icons intuitive is self-evident.

Consider the following pair of alternative graphics, each of which is intended to depict the removal of a drug patch's backing liner.

The graphic on the left represents the user's intended action in a more abstract manner, using the combination of the arrow and trash can symbols to connote the liner's removal. The illustration on the right uses a more representative approach, showing the specific movement users must make with their hands to remove the liner from the patch.

While the intended meaning of the graphic on the left was undoubtedly clear to its illustrator, we suspect that its meaning might be obtuse to many users. Furthermore, some users might misinterpret the graphic and throw away the patch rather than remove its liner. On the other hand, the more representative illustration on the right is likely to be clear to most users.

As you can see, such intuitive graphics do not depend on words to convey meaning. Most people glancing at a well-designed graphic will understand its meaning. Some details in these graphics might be unclear to some people, but few, if any, are likely to misinterpret them in a critical way.

Intuitive illustrations tend to have one or more of the following characteristics:

The product and its accessories are depicted in a representative rather than abstract manner.

Drawings depict elements using a visual style and perspective that foster rapid recognition.

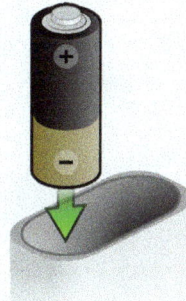

Elements are illustrated in a consistent manner rather than varying widely in appearance.

Representative and appropriate colors are used when color is critical to interpreting the shown element (e.g., red stop sign).

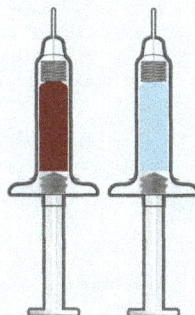

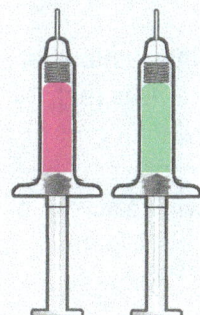

Graphical elements have simple form and limited, but salient details that foster understanding.

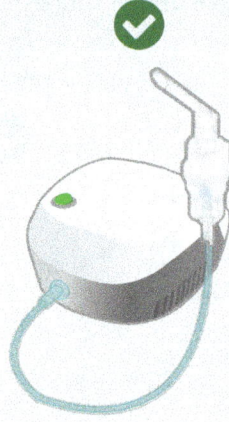

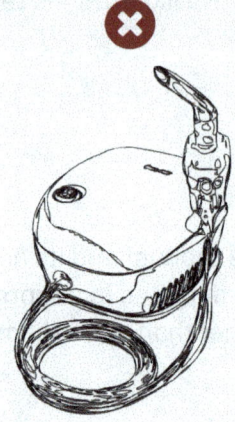

Elements provide necessary context, such as bubbles floating around hands being cleaned.

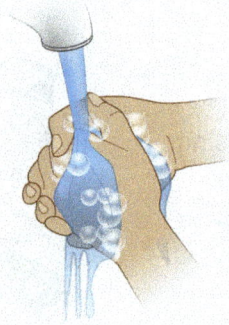

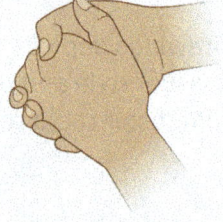

Capture the "scene" at its most informative moment, typically immediately before or while the instructed action occurs.

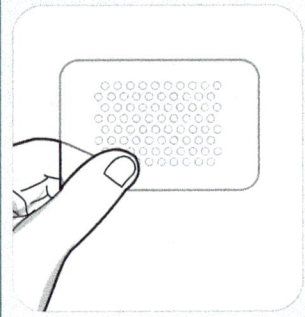

The interaction between multiple elements is realistic, such as showing fingers gripping the barrel of a pen-injector and showing a screw-on needle assembly in the proper order.

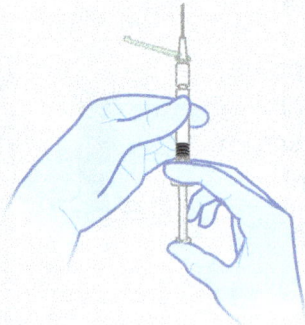

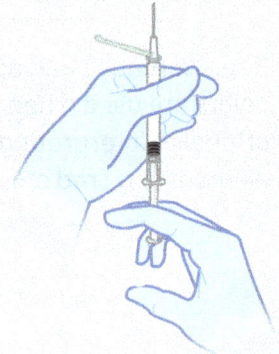

These suggestions are mostly applicable to illustrations that depict interactions with the product in a relatively representative way. But sometimes utilizing a more abstract icon or symbol might be a more effective means of visual communication.

Consider using icons or symbols when:

- Conveying simple, universal concepts, such as power on/off, battery, or temperature

- Space is limited and a simple graphic would be more effective than text alone

- Needing to reinforce repetitive information, such as reminders or warnings

- Enhancing quick recognition by using familiar symbols to reinforce the step, such as with a power-on, Wi-Fi, or Bluetooth symbol

SIDEBAR
Using standardized symbols

There are advantages to using icons and symbols with which users are already familiar. For one, users are more likely to recognize them and interpret them correctly. Second, regulators, including the FDA, encourage manufacturers to utilize standard symbols whenever possible, rather than create novel designs unnecessarily.[*]

Fortunately, the international standard ISO 15223-1 *Medical Devices - Symbols to be used with information to be supplied by the manufacturer* provides a collection of symbols for depicting common content in medical device instructions.

Examples of symbols included in ISO 15223-1.

---

[*]   Title 21, Code of Federal Regulations, Section 801.15.

53

Make illustrations clear

The best illustration is one that immediately communicates its intended meaning with little if any ambiguity. You could say the meaning of the illustration is self-evident.

In general, illustrations tend to be clearer and easier to understand when they adhere to the following guidelines:

Make visual elements recognizable, matching the real items, and make them reasonably representational without being too detailed.

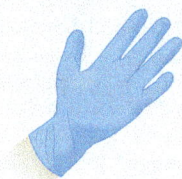

Make the illustration large enough that the user does not overlook key parts.

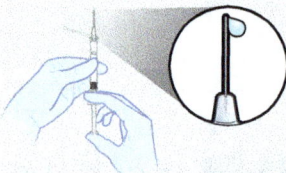

Use bold lines to make the major visual elements appear unified.

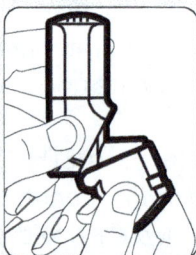

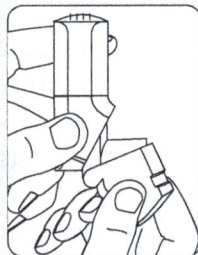

Use solid color or gradient fills to give integrated elements more form and avoid making parts look truncated.

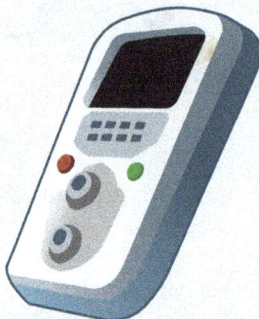

Depict a moment in time and, where necessary to ensure understanding, create separate illustrations showing the beginning and end states of a given task.

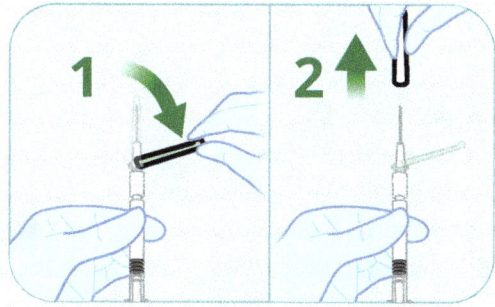

Where necessary to communicate things lacking a strong physical analog (e.g., "wait 5 seconds"), add a limited number of symbols, text, and words to amplify the graphical information, but strive to design illustrations that do not depend on these additions.

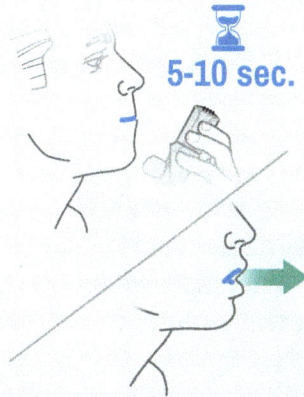

Consistently depict actions as the reader might observe themselves performing the task (i.e., first person view) or someone else performing the task (i.e., second person view).

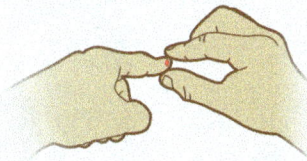

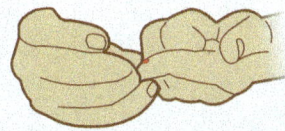

Draw attention to key (i.e., salient) details.

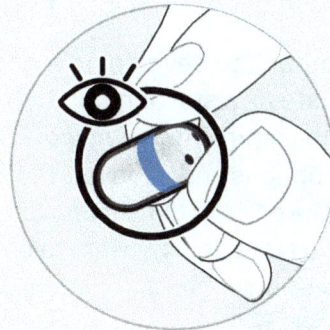

Subordinate or, better yet, exclude extraneous details.

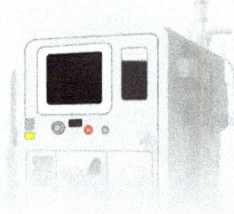

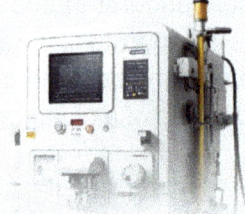

Reserve distinct and salient colors for key details in the illustration, rather than using them for decorative purposes.

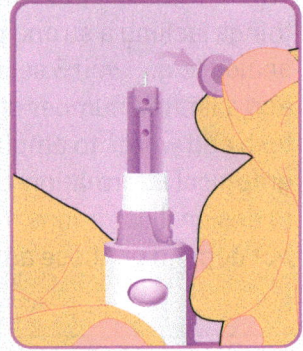

Add accent lines sparingly but where necessary to add clarity.

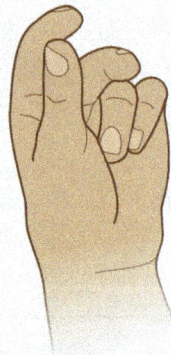

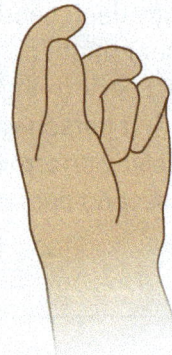

A proven approach to developing a good illustration is to clarify what you are trying to communicate as a starting point. Then, conduct an image search for comparable illustrations to inspire your own work while respecting copyrights and the like. Next, sketch many options and choose a few of the best ones. Now is the time to collect user feedback on the options to help converge on the best one. You can then develop a final illustration and retest it with users to ensure that it is effective (see "Verify and validate the instructions" on page 73).

Evaluating illustration clarity

There are a few ways to test the clarity of an illustration.

View at-a-glance

Present the illustration on a card or a screen for around 5 seconds, remove it from the view, and then ask the user to state its meaning. This approach is arguably the most rigorous because the given illustration stands alone without context. One ANSI standard[*] on warning symbols suggests that users should get the meaning correct at least 85% of the time and that critical confusions should not exceed 5%.

View as a set

Present several illustrations at once on a card or a screen. While the illustrations remain in view, ask the user to state the meaning of each one. This approach allows users to decipher the meaning of each illustration based on its own characteristics but also in view of the other illustrations that create context from which users can derive deeper meaning.

Match to definitions

Present several illustrations on a card or a screen along with a randomized list of meanings. While the illustrations remain in view, ask the user to match each illustration to one of the listed meanings. This is a relatively weak means to evaluate an illustration but expedient.

A usability test participant views illustrations on a laptop display (left) then explains his interpretation of the illustrations (right).

---

[*]   ANSI Z535.3-2022: Criteria for Safety Symbols. American National Standards Institute. 2022.

54

Direct attention to important details

Even relatively simple products that consist of only a few components can seem intimidating to those who haven't used them before. With more complex products that feature myriad controls, components, and accessories, users are even more likely to be overwhelmed. Instructions can help calm users and guide them in the right direction by highlighting important details in the product or its use.

Nature provides countless examples of flora and fauna that draw attention to themselves using colors, shapes, and spacing. Yellow tulips rely on their vibrant colors to entice pollinators. Adult penguins have contrasting feather colors that help them stand out in a creche of baby penguins. Male fiddler crabs have one oversized claw that sets them apart from other crabs. The carnivorous sundew plants draw attention to themselves with sticky tentacles arranged in a spaced-out radial pattern.

Examples of attention-drawing methods found in nature.

People have adopted similar ways to draw attention to themselves, their surroundings, or their artifacts, such as by making something larger than surrounding items, placing it in a conspicuous location, creating space around it, and placing it in a special setting or against a special background.

Examples of attention-drawing methods used by people.

You can use similar techniques to draw attention to important information, including text and graphics in instructions.

You can emphasize a key element of a graphic by:

Making it relatively large

Step 2
Press the
POWER button

Placing it at the center
of the graphic

Separating it from the rest
of the composition

2 Insert syringe into heparin port.

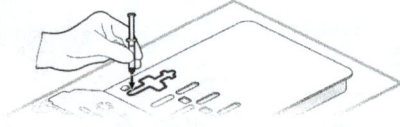

DO NOT infuse
heparin into port.

Presenting it in an
interesting context

REMEMBER
**Check for
air bubbles**

Pointing to the key element

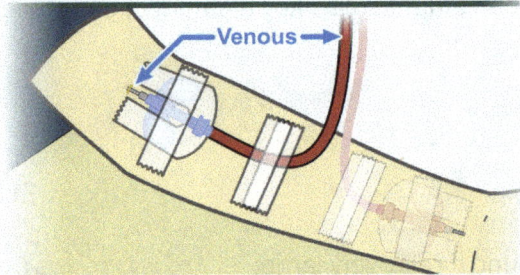

←Venous→

Secure the venous **needle** and **line** to the
your clinic's protocol.

Outlining it with an
enclosing shape

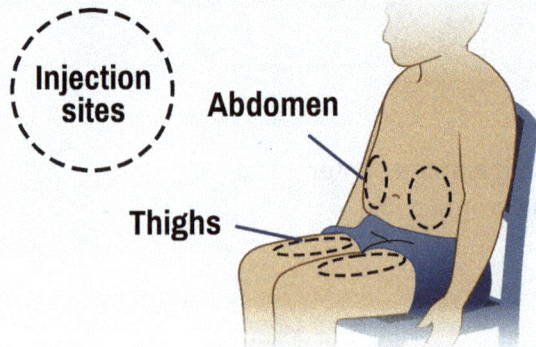

Injection
sites

Abdomen

Thighs

Desaturating other
elements in the graphic

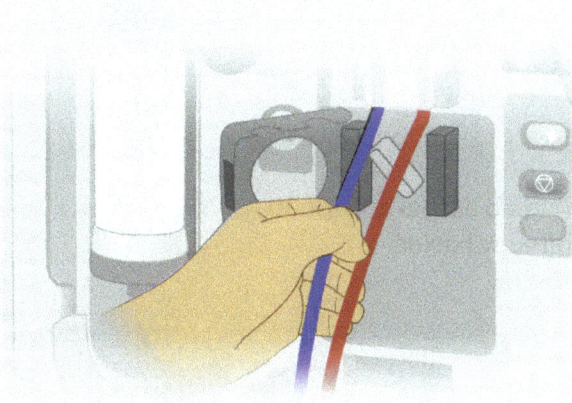

Coloring it with a
contrasting palette

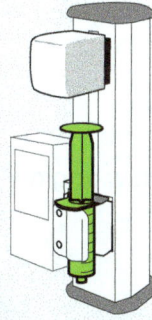

Place **syringe** in holder.

Drawing it using bolder lines
than other elements

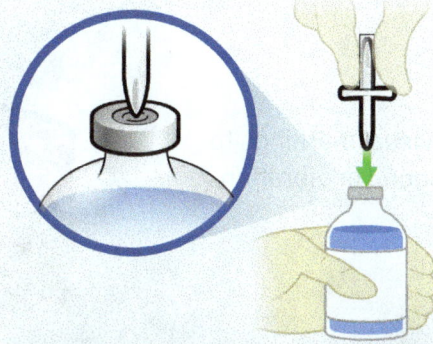

Using a "spotlight"
effect to highlight it

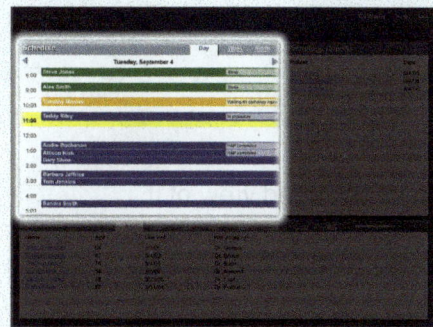

**The schedule is displayed in the
screen's upper left.**

Adding a label or caption directly beside it

5 seconds

Adding a drop shadow or glow

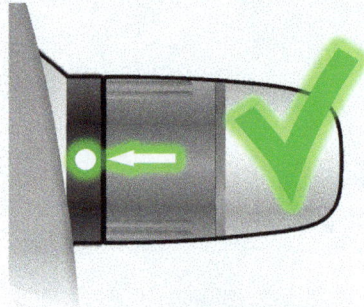

Locked:
Arrow and dot aligned

Adding visual markings to depict sound or vibration

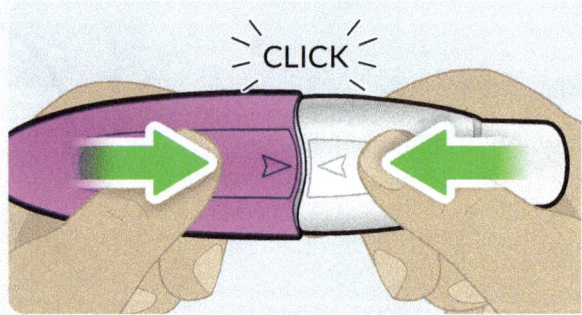

CLICK

Including an enlarged view of it

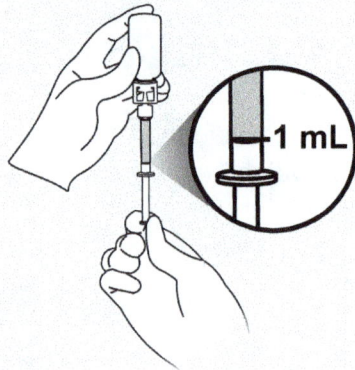

1 mL

When drawing attention to text, you can emphasize a specific word, sentence, or paragraph by:

Using a bold, italicized, or underlined font	**WARNING** <u>ATTENTION</u> *NOTICE*
Using a larger font size	Need help? Our knowledgeable staff is available to provide you with guidance, troubleshooting support, and answers to your inquiries to ensure accurate and effective use of your device. Contact us by phone at
Coloring it with a contrasting or eye-catching color	**To end therapy, press the Stop button.**
Capitalizing all letters (i.e., using all-caps)	DO NOT attempt to repair the device yourself.
Highlighting it using a background color	**REMEMBER** **Check battery before use**
Enclosing it in parentheses or brackets	To keep changes to the EHR, click [Save].
Indenting it to separate it from other content	3. Attach the electrode pads to the patient's skin, making sure it is adhered securely. IMPORTANT: Do not place pads over broken skin or sensitive areas.
Oversizing the initial letter (i.e., using drop-caps)	Before you begin, read all instructions closely.

Importantly, for these techniques to work, you should use them sparingly. If you overuse them, you'll draw attention to many elements in the document, thereby canceling out the techniques' intended effects.

55
Show the product

Ask yourself this question: do you want to see a picture of the pizza contained in a frozen pizza box? Or would it be okay for the box just to be labeled "Pizza?"

We're pretty sure you want to see the pizza depicted in its full, reheated glory, right? It tells you what you are getting, albeit with a degree of visual embellishment. Good or bad taste notwithstanding, the pizzas never look that good after you reheat them.

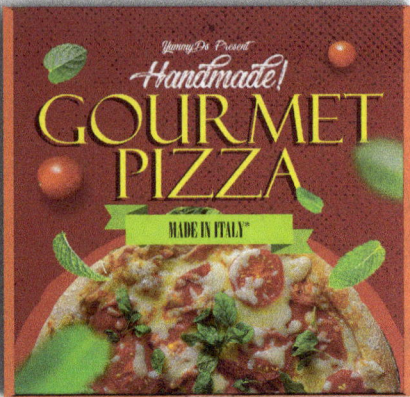

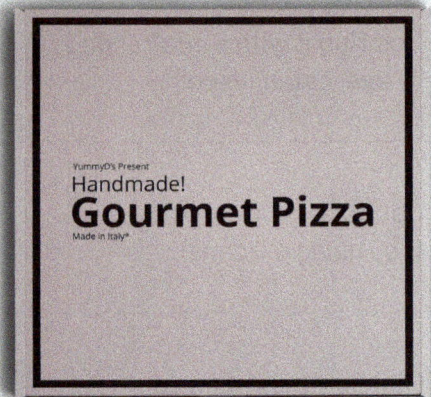

For the same fundamental reason that it helps to show a pizza on its box or a medical device on its packaging, it's smart for instructions to include a picture (or drawing) of the associated medical device near the start of the instructions. It satisfies an emotional need to see that the instructions are matched to the device.

Logically, you should place the product's image close to its trade name, which regulators expect to see in the instructions, particularly for drug-delivery devices (i.e., combination products).

Make sure the image is sufficiently accurate because inaccuracies might cause readers to worry they have the wrong instructions. That said, it should be okay to limit the amount of visual detail in a drawing to the point that the device is recognizable and includes all salient elements that might distinguish it from another device.

What if there is a lot of content to include in the instructions and space is tight? Is it okay to leave out an image of the medical device? It depends on the product type. For example, the FDA requires all combination products released in the USA to include an image of the product in the instructions for use. For many medical devices, including images of the product in the instructions will be critical to the instructions' effectiveness. However, even in cases where including an image isn't required for regulatory reasons, we are pretty sure users will be more satisfied if you can include the image without visually congesting the rest of the page.

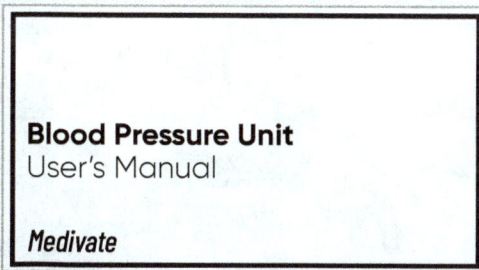

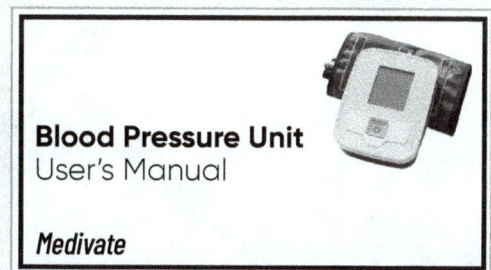

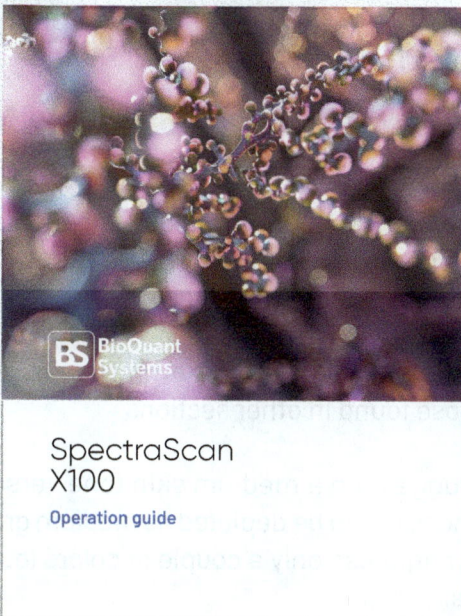

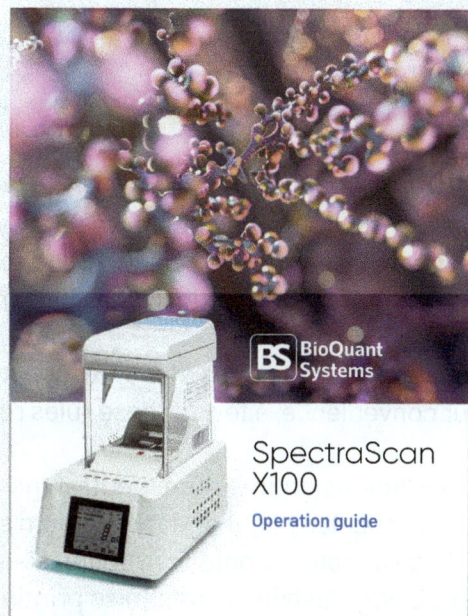

Examples of document covers with and without an image of the associated product.

56
Drawing hands

Considering the bone structure, musculature, surface features, and a seemingly infinite number of possible hand positions to depict, drawing the human hand can be quite challenging. The task appears all the more daunting when looking at examples from art history.

A detail from "The Creation of Adam", part of Michelangelo's Sistine Chapel fresco.

To put it mildly, Michelangelo was pretty good at hands. Luckily, you don't need to be that good. In fact, you just need a good reference photo and know how to use a drawing application, (see "Creating illustrations from reference photos" on page 176).

Still, we offer these rules of thumb (pun intended) to consider when drawing hands. For your convenience, a few of these rules recap those found in other sections.

- Depict skin in a race-neutral manner, suggesting a medium skin tone versus particularly light or dark. Medium skin tone can also be depicted as medium gray on a monochrome instruction sheet that might use only a couple of colors (e.g., black and blue) to minimize printing costs.

- Depict hands in a sex/gender-neutral manner, suggesting a medium-sized hand. The exception would be when depicting use of a product that is only used by a person of a particular sex/gender.

- Make the hands look more recognizable by adding just a few details, such as knuckle wrinkles, fingernails, and even the "half-moon" within a fingernail. Such details provide subtle but helpful cues about finger positioning and orientation.

- Consider depicting overlapping fingers, which can add realism to some hand drawings by providing a three-dimensional appearance that can help indicate hand position relative to the product's orientation.

- Use extra shading and shadows to add pleasing style and a subtle three-dimensionality to hand drawings, which also helps convey hand posture more effectively.

- Depict realistic hand postures to the extent possible without blocking essential details of other objects.

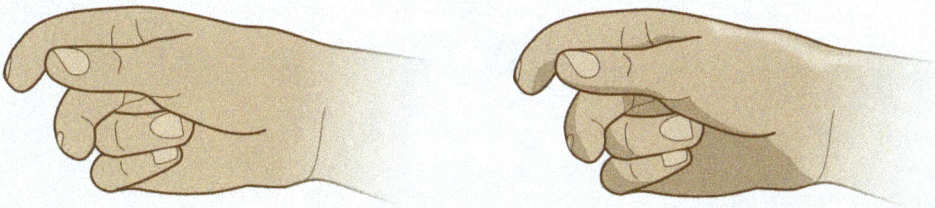

- Show hands wearing protective exam gloves when appropriate.

- Consider whether you can communicate an action better by showing its starting point or endpoint. For example, you can show a user's hand dialing 30 units of

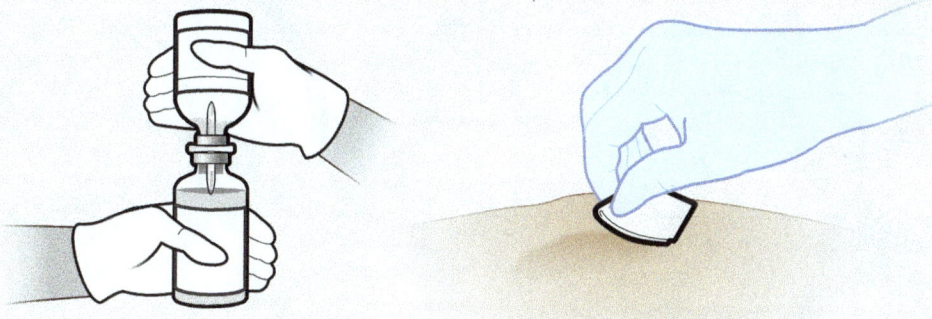

insulin on a pen-injector by showing a hand that has already twisted a dial and displaying 30 in the dose window — the endpoint. Alternatively, you can illustrate the user holding a pen-injector's cap in one hand and the pen-injector in the other hand, plus an arrowhead indicating to attach the cap.

- Add movement lines and possibly ghost images to convey motion.

- If arm position is not relevant, truncate hand drawings below the thumb or at the wrist crease.

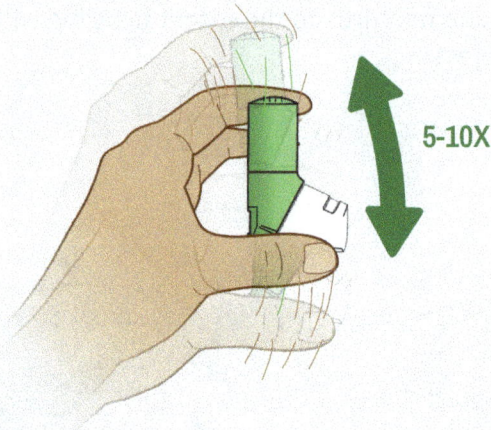

- When possible, show the hands as they would appear from the user's perspective. This might help users mimic the depicted action more effectively by utilizing their mirror neurons — brain cells thought to help people imitate observed behaviors.

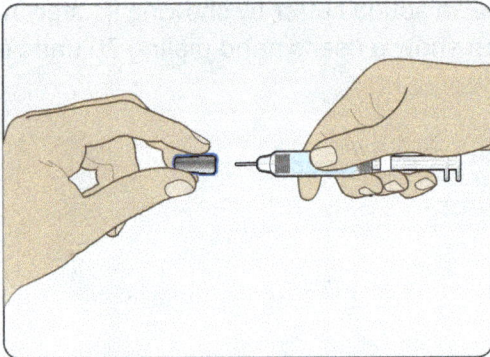

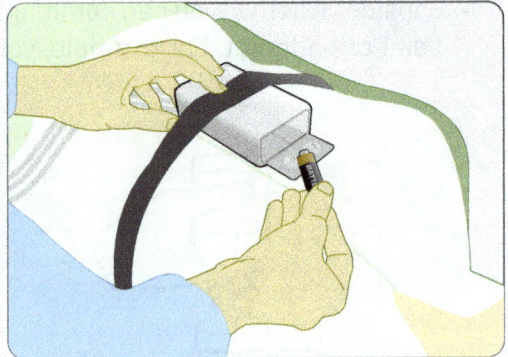

- Show a finger alone if that is enough to clearly convey the intended action (e.g., a button press). Conversely, avoid "zooming in" on a small part of a finger or hand and not giving enough context regarding the intended position or posture.

- Show slightly exaggerated hand positions to depict hand positions that require extra force or precision (e.g., pinch or power grip).

- Make sure that all illustrations in the document depict hands from a consistent point of view with a consistent illustration style and level of detail.

- Do not show the hand if it adds no instructional value to a given graphic.

- Visualize a smell, feel, or sound in a widely recognized, conventional manner such as diverging lines to suggest a "click" feel and sound.

SIDEBAR
Why do cartoon characters wear gloves?

You might have noticed that many cartoon characters, such as Mickey Mouse, Bugs Bunny, and Woody Woodpecker, wear gloves. This isn't because they're practicing aseptic technique. Rather, the gloves help bring attention to the characters' hands by dressing them in a color that contrasts with the characters' bodies and other elements in a scene.[*] This is crucial for animators, because hands provide a key way to express a character's personality.

This principle is useful for instructional illustrations as well. Although not all medical product instructions require the depiction of gloves, the hands should always stand out from other parts of the illustration to help users perceive them. You can do this by depicting hands in a tone or color that contrasts sufficiently with other parts of the illustration.

On the other hand (pun intended), some practices are best left for cartoons, such as the tendency to draw hands with only four fingers. Reportedly, animators do this to reduce the effort associated with animating hands. For instructional illustrations, we suggest investing the extra effort in creating anatomically accurate five-fingered hands.

---

[*]   Emily DiNuzzo, "This Is Why Most Disney Characters Wear Gloves" Reader's Digest, March 30, 2023, https://www.rd.com/article/why-do-disney-characters-wear-gloves/

57

Creating illustrations from reference photos

Arguably, the optimal way to create illustrations is to hire a professional illustrator who specializes in creating clear technical illustrations. In lieu of that option, you can create effective illustrations by photographing the actions you want to depict, then trace the photos using vector-based illustration software, such as Adobe Illustrator, CorelDraw, or Affinity Designer.

The advantage of vector-based illustrations is that you can scale them — make them larger or smaller — and they will remain smooth and sharp rather than pixelated or blurry. Tracing photographs in an illustration program is a matter of clicking on parts of the image to form "anchor points" and then dragging on "handles" to form curved lines.

It is good practice to draw complete and separate objects. For example, in the illustration below the objects would be the syringe, the patient's arm, each glove, and the tourniquet. Then, you can layer, scale, rotate, and modify these elements as needed to create the desired scene. However, you might also need to create several versions of a given object to depict it from different angles or in different states (e.g., different hand positions).

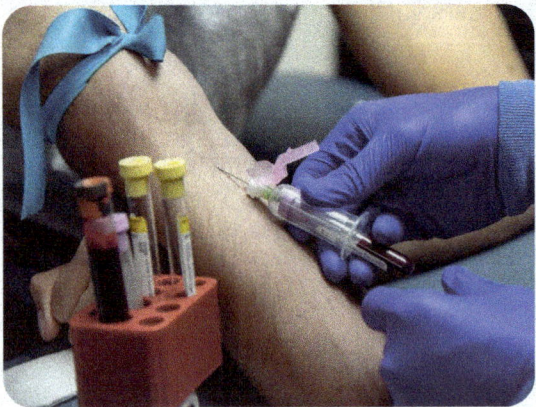

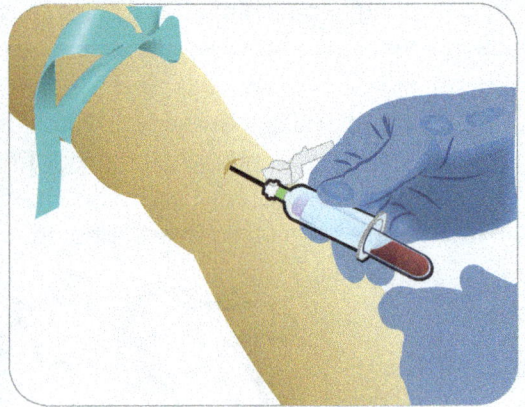

A reference photograph (left) and the resulting vector-based illustration (right).

CHAPTER SIX
PRODUCTION CONSIDERATIONS

58

Paper thickness

Beware of the thin paper problem.

Instructions printed on thin paper that lacks good opacity can suffer from show-through (i.e., "bleed through"). That is to say, when viewing one side of the document, you can see some of the reverse side's content. This effect reduces the content's color contrast and creates visual noise, thereby reducing legibility and readability.

See what we mean:

IFUs printed on thin paper that lacks good opacity can suffer from show-through (i.e., aka "bleed through"). That is to say, when viewing one side of the document you can see some of the reverse side's content. you can see some of Side 2's content when reading Side 1's content, and vice versa

There are two basic ways to reduce or eliminate the possibility of show-through, both related to the opacity of the paper stock. One way is to print on relatively thick paper. However, using thicker paper can drive up costs and might complicate folding. Another way is to use an uncoated paper. Typically, uncoated papers have a higher pulp content (and less opacity) than coated papers with a lower pulp content (and more opacity).

If production or budget constraints prevent you from using thicker or uncoated paper, the next best option is to arrange content on the two sides in a manner that minimizes overlap. For example, text-heavy areas on one side of the document should have a reverse area with relatively low information density, thereby diminishing the show-through effect.

Commonly used
paper thicknesses

in gsm (Grams per Square Meter) and lb (Pounds)

12-16 gsm (8-10 lb)
Extremely thin paper often used for
toilet paper and other similar products.

20-30 gsm (12-20 lb)
Often used for tissue paper, like the type
found in gift bags or shoe boxes.

35-55 gsm (24-37 lb)
Used for lightweight paper products, such
as newsprint or thin paper used in many
pharmaceutical instructions.

75 gsm (50 lb)
Standard copy paper or printer paper
thickness.

90 gsm (60 lb)
Slightly sturdier and higher quality than
standard printer paper; used for resumes
and formal documents.

89-105 gsm (60-70 lb)
Often used for flyers, leaflets, and posters.

105-120 gsm (70-80 lb)
Often used in higher end notebooks and
sketch pads.

120-145 gsm (80-100 lb)
Commonly used for magazine covers,
paperback book covers, and brochures.

145-185 gsm (100-120 lb)
Used for high-end greeting cards and
business cards.

59
Fold lines

Fold lines — formed where a sheet is folded one or more times – might seem like a non-critical aspect of an instructional document's design. But you'd likely be surprised by the effect they have on users' perception of the document's contents.

In some ways, fold lines on instructions act like the demarcation lines found on control panels.

On this panel, white lines serve to demarcate functionally related groups of controls and indicators.

Demarcation lines — the white lines in the above control panel — create groupings that help users interpret the purpose of individual controls and displays, and how they do or do not relate to the surrounding groups. Often, such functional groupings feature a descriptive label as well. For more detailed information on demarcation lines and their use with regard to instructional material, see "Use demarcation lines" on page 114.

It doesn't matter that the fold lines on a document are not actually printed lines. Folding a document for packaging divides it into rectangles that stand out when it is unfolded due to both the fold lines themselves and the effect of light on the folded page. This sets the stage for users to view the content one rectangle at a time, and sometimes as a contiguous row or column comprised of multiple rectangles. Therefore, take advantage of this user perception and avoid violating the users' expectations.

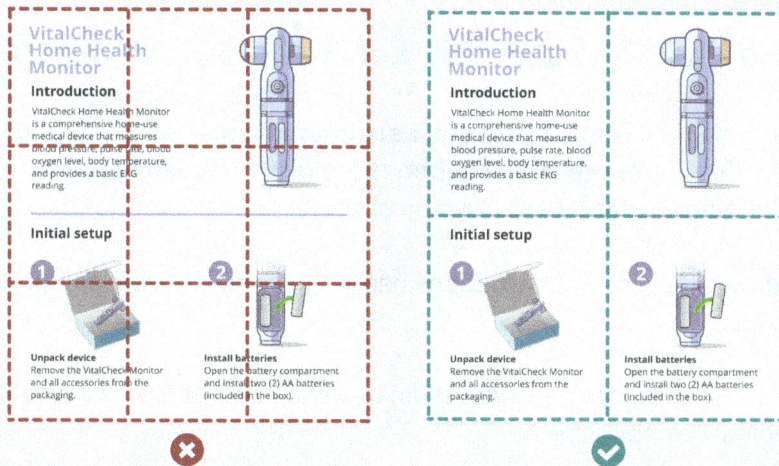

Above, you can see the difference between content that is enhanced versus disrupted by fold lines (represented by the dashed lines).

Here are some design tips for utilizing folding lines effectively:

- If needed, spread related content, such as a paragraph of text, across multiple rectangles that you can treat as a single, integrated block.

- With some languages, users are likely to read blocks of content left-to-right (i.e., one row at a time) rather than vertically. With other languages, the reading pattern might be column by column and start on the left or right side. So, be mindful of the document's language when choosing the order of content defined by the fold lines and any other directional cues.

- Minimize the number of folds to the extent that the packaging allows.

- Avoid mixing unrelated content in the same rectangle (i.e., section) or perceived set of rectangles.

- Do not print rectilinear content (i.e., straight lines) on a fold line because (1) the fold might interfere with the content's legibility, and (2) the line might not align perfectly with the fold, in which case the user will effectively see two demarcations instead of one.

- Avoid fold patterns that create odd rectangle aspect ratios and make content blocks appear too narrow or wide relative to the overall composition.

60

Draw attention to the instructions

Just because you have produced great instructions does not mean that people will look at them. This is a real problem when there is critical information to convey to the user and the instructions are the only channel of communication.

The need to draw attention to instructions has led to some creative solutions. Here are some good ones:

- Add a label to the outer package telling users to read the instructions before using the medical device within.

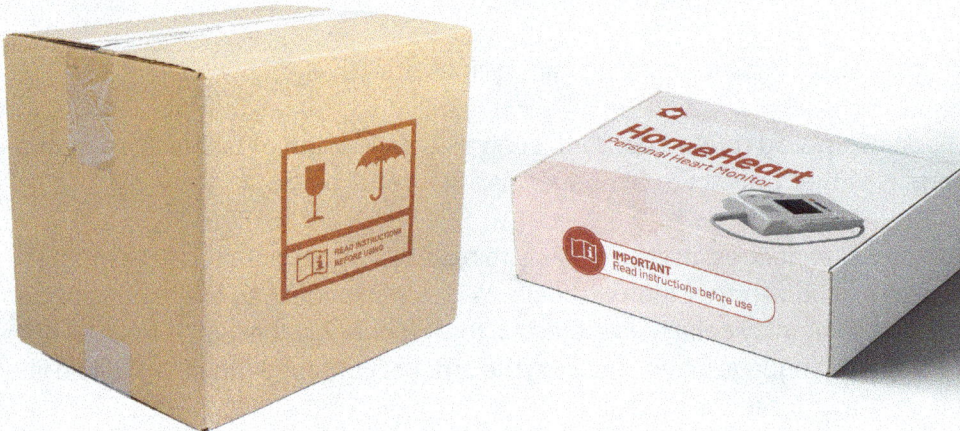

Examples of medical device packages that prompt users to read the product's instructions before using the product.

- Make the instructions stand out visually, such as by making the outer panel an attention-getting color.

- Write "instructions for use," "read me first," "quick reference guide," or the equivalent to distinguish the instructions from other documents (e.g., warranty information) found in the package. For some products, such as drug-delivery devices, labeling the instructions as "instructions for use" in bold, capital letters is recommended by the FDA.

- Place the instructions (usually in a compact, folded form) so that users must encounter them first when opening the package containing the medical device. In other words, deliberately make it an obstacle to device use. Because package size is usually minimized to reduce waste and storage requirements, the instructions might need to be wrapped around the medical device, such as a metered dose inhaler. You could even tape the instructions on to the medical device in a place that demands attention before people use the device. For example, over the device's power button.

Example of an instructional document that is visible immediately when the user opens the product's packaging.

- If the medical device has a computer display, use it to prompt people to read the instructions, perhaps even requiring them to acknowledge the prompt. This might seem like a relatively weak mitigation against the risk of a user ignoring the instructions, but it's something in lieu of nothing.

- Ensure that medical device training sessions delivered in-person, remotely, and online include reviewing the instructions.

CHAPTER SEVEN
EXEMPLARS

Disclaimer
This chapter presents <u>hypothetical</u> instructions that showcase design characteristics advocated throughout the book. The exemplars do not depict actual products, nor do they represent real-world instructional materials.

While we emphasize the importance of conducting thorough user research and usability testing to create high-quality, user-centered instructions, the exemplars in the following pages are not the result of such research. Instead, they are informed by our collective experiences working on instructions for products similar to the hypothetical ones depicted here.

As such, these exemplars are intended for illustrative purposes only. They are not meant to serve as templates or to be directly applied to real-world instructions you develop.

IFU for a combination product

Product

Transdermal medication patch - a combination product consisting of a drug and a medical device, which delivers medication continuously when placed on the patient's skin.

Target users and use environment

Patients and caregivers who use the product at home or in other non-clinical environments.

Document format

A single A3 sheet (297 x 420mm) folded into 9 panels with black ink printing on both sides.

Notable characteristics

- **Compliance with regulatory guidance** - As an IFU for a combination product, the document complies with FDA CDER's guidelines for content and formatting of combination product instructions (see "Review regulatory guidance" on page 50)

- **Standard paper size** - The IFU is printed on an A3 sheet, which facilitates and lowers the cost of printing and packaging (see "Select orientation (aspect ratio)" on page 84)

- **Continues on back** - The lower right corner of the IFU includes a prominent indication that additional content is available on the back side (see "Continue on back" on page 115)

- **Harmonized fold lines** - The IFU's grid-based layout aligns with the 9-panel folding pattern, ensuring that the fold lines do not affect the document's legibility (see "Fold lines" on page 182)

- **Sufficient paper thickness** - Although the IFU is printed with black ink, the use of sufficiently thick paper prevents ink bleed-through (see "Paper thickness" on page 180)

...TIONS FOR USE
...RMAPREL
...ide transdermal system)

These Instructions for Use contain information on how to apply, remove, and dispose of DERMAPREL transdermal system (patch).

...ctions for Use before you ...ERMAPREL and each time

Document issued: June, 23 2024
PN: 31298636l6AOHH
© 2024, VIREXA, INC.
All rights reserved.

Important information to know before applying DERMAPREL

1. Avoid placing the patch on irritated, broken, or damaged skin, as this may affect absorption or cause discomfort.
2. Consult your healthcare provider if you have a history of skin sensitivities, as some users may experience irritation or allergic reactions.
3. Keep the patch away from heat sources like heating pads, electric blankets, or direct sunlight, as this can increase medication release and lead to overdose.
4. Do not apply more than one patch at a time, unless directed by your healthcare provider.
5. If you experience dizziness, nausea, or difficulty breathing while using the patch, remove it immediately and seek medical attention.

System contents

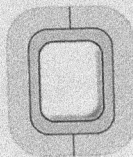

Dermaprel patch (x7)   Instructions for Use (this document)

Required accessories (not included)

Saline wipes   Gauze pads

...g a new patch

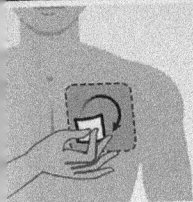

...ve patch
...n hair on your left chest to ...ive patch will contact skin.
...ng saline pad.
...area with gauze to dry.

2

Remove patch liners
a. Peel off white liners numbered 1 and 2 from adhesive patch.

3

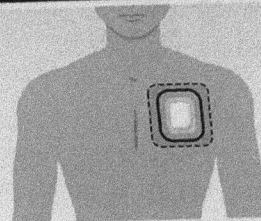

Adhere patch to left chest
a. Make sure skin is completely dry before placing patch.
b. Avoid stretching patch when applying it to the skin.
c. Place patch's adhesive side on left chest then firmly press it against the skin to secure it.

...ng a used patch

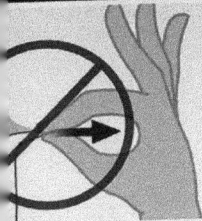

...the adhesive patch
...ar from the skin. Doing so will
...risk of a skin injury.
...d or flex the Shield sensor.
...sensor can damage it and
...om communicating with your
...ce.

1

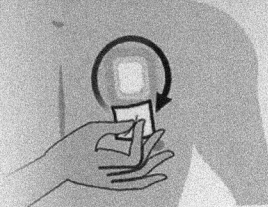

Loosen adhesive patch
a. Apply warm water or baby oil around the edges of the adhesive patch.
b. Continue applying water or baby oil until the edges of the adhesive patch begin to lift.

2

Slowly lift and remove adhesive patch
a. Begin peeling off the adhesive patch slowly at a low angle. With your other hand, press down the skin under the lifted patch.
b. Continue pulling patch slowly away from the skin and pressing skin down with your other hand until you fully remove the patch.

CONTINUES ON BACK ≫

Quick reference guide for an over-the-counter medical device

Product

Peak flow meter (also known as a spirometer) - an over-the-counter device used to measure a patient's lung capacity.

Target users and use environment

Patients, including those with asthma or other respiratory conditions, and caregivers using the device at home or in outpatient settings to monitor lung function.

Document format

A single letter-sized (8.5 x 11 inches) sheet, printed in full color on both sides, folded into a tri-fold format.

Notable characteristics

- **Consistency with product design** - The document's color palette aligns with the device's industrial design, creating a pleasing harmony between the instructions and the product (see "Instructions should harmonize with the product" on page 17)

- **Prominent step numbers** - Each step is marked by a large, highly visible number, making it easy for users to follow the correct sequence of actions when using the device (see "Take it step-by-step" on page 90)

- **Simple composition** - The horizontally arranged steps follow the natural reading flow, making the instructions intuitive and easy to read (see "Use composition principles" on page 111)

- **Low information density** - The high ratio of white space to content creates a simple and pleasing visual design (see "Make instructions visually appealing" on page 117)

- **Compact format** - The tri-fold layout ensures the document is compact enough to fit into the device's pouch, offering both convenient storage and easy access (see "Draw attention to the instructions" on page 184)

Check bat
to make su
meter conta
enough pow
open purple

w to measure peak flow

2

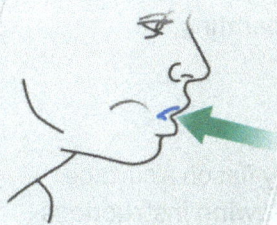

Breathe in as fully as you comfortably can through your mouth.

3

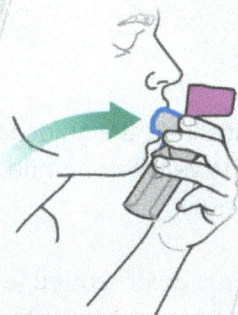

Insert mouthpiece into your mouth, keeping your lips sealed around it.

Blow out through your mouth as strong and fast as you can in a single blow.

4

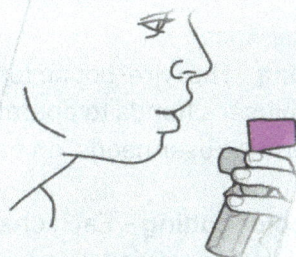

Remove meter from mouth, then continue breathing as usual.

5

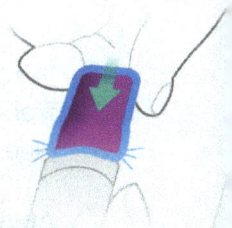

Close purple cap to prepare for your next measurement.

⚠️

IMPORTANT
You must close cap **after every measurement.**

If you do not close the purple cap, the meter will not count your breaths correctly.

If
inc
one
step
addit

User manual for a prescribed home-use medical device

Product
Home-use ventilator - a medical device intended to assist patients who require respiratory support.

Target users and use environment
Patients and caregivers using the ventilator in non-clinical environments, such as at home.

Document format
A letter sized, wire-bound booklet featuring color-coded tabs separating the guide's primary sections.

Notable characteristics

- **Wire binding** - The wire-bound format allows the guide to lay flat on a surface, freeing the user's hands to operate the ventilator while following instructions (see "Determine user needs" on page 40)

- **Chapter color coding** - Each chapter is color-coded (e.g., Daily Maintenance in green), helping users visually associate specific layouts with their corresponding sections ("Use color effectively" on page 100)

- **Conspicuous warning** - The caution message is displayed directly next to the relevant procedural step and is highlighted with a bold yellow background to ensure visibility and user compliance (see "Warn users about potential risks" on page 106)

- **Color-coded tabs** - Tabs with distinct colors enable users to quickly access each chapter of the guide, improving navigation and ease of use (see "Help users find content of interest" on page 121)

- **Highlighting salient details** - Thick black outlines in the illustrations emphasize the components users need to interact with at each step, drawing attention to critical parts of the process (see "Direct attention to important details" on page 164)

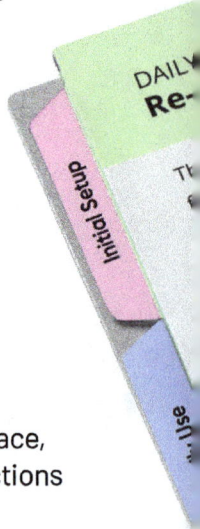

e water tank

...n your ventilator helps keep the air moist and comfortable. It's important to
...r it regularly so your ventilator works well. This page shows you how to
...nd put back the tank safely.

Unlatch water tank
Press the release latch or button to unlock the water tank.

2 Remove water tank
Gently pull tank out of its compartment, taking care not to spill water.

3 Fill water tank
Fill the reservoir with distilled water to the "Max" fill line.

⚠ **CAUTION: Only use distilled water**
Tap or mineral water can damage the device.

4 Insert water tank
Hold the water tank above its slot, making sure to align the tank's bottom drain with the circular port.

5 Secure water tank
Gently push the tank into the ventilator to secure it.

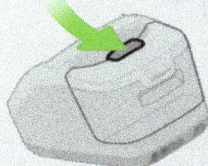

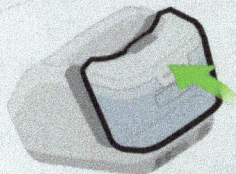

Replacing the air filter

The air filter in your home ventilator helps kee...
particles. Replacing the filter regularly is imp...
and provide clean air.

1 Open air filter compartme...
Press the release latch or ...
open the air filter comp...

2 Remove old filt...
Gently pull ou...

3 Th...

Quick reference cards for a medical device used by clinicians

Product
Infusion pump - a medical device used to deliver fluids, such as medications or nutrients, into a patient's body.

Target users and use environment
Healthcare professionals, such as nurses and clinicians, who need to operate the infusion pump efficiently and safely in hospital wards and intensive care units.

Document format
A set of small (5 x 8 inch), laminated cards bound by a ring.

Notable characteristics

- **Alignment with user needs** – Each card is tailored to address specific tasks users commonly encounter while operating the pump, providing targeted assistance exactly when and where it's needed (see "Determine user needs" on page 40)

- **Compact and durable format** - The cards are relatively small and laminated, making them easy to handle and resistant to damage from frequent use in clinical environments (see "Determine production opportunities and constraints" on page 57)

- **Highlighting key details** - Important terms, such as button labels, are bolded throughout the instructions, allowing users to quickly scan and locate essential information without needing to read the text in full (see "Sizing and styling text" on page 124)

- **Use of screenshots** – Images of the pump's software user interface are included alongside the instructional text, helping users quickly identify buttons and controls on the pump's display (see "Show the product" on page 170)

- **Ring-bound document** - The cards are bound by a metal ring, enabling them to be hung directly on an infusion pump, making them readily accessible during operation (see "Draw attention to the instructions" on page 184)

Program a basic infusion

1	Press **Rate**	Rate ── mL / hr
2	Enter a rate, then press **Confirm**	50 mL / hr keypad: 1 2 3 / 4 5 6 / 7 8 9 / . 0 Del — Cancel / Confirm
3	Press **VTBI** (Volume to be infused)	VTBI ── mL
4	Enter a volume, then press **Confirm**	250 mL keypad: 1 2 3 / 4 5 6 / 7 8 9 / . 0 Del — Cancel / Confirm
5	To begin the infusion, press **Start**	Rate 50 mL/hr / VTBI 250 / ▶ Start

CHAPTER EIGHT
RESOURCES

Design checklists

To help you apply our guidance to real-world projects, here are two checklists based on the book's content.

The first outlines key steps for creating instructions and can serve as an initial roadmap for your instruction development efforts. The second presents key design guidelines to help you evaluate your instructions.

Process checklist

Planning and preparation

☐ Identify the product's and instructions' users.

☐ Conduct user research to identify user needs and expectations for the instructions.

☐ Document user needs, preferences, and contexts of use.

☐ Identify and assess the product's use-related risks.

☐ Review applicable regulatory requirements and guidelines for the instructions (e.g., FDA, EU MDR).

☐ Establish the instructions' objectives and scope.

☐ Define the key functions and tasks that the instructions must cover.

☐ Establish the instructions' requirements.

☐ Determine needs for translation and localization.

☐ Identify production constraints and opportunities.

Initial design

☐ Draft an initial outline for the instructions' content and structure.

☐ Develop alternative conceptual models for organizing and presenting information in the instructions.

☐ Consider and explore the use of alternative media options (e.g., printed booklet, quick reference guide, e-IFU).

☐ Create preliminary sketches (digital or hand-drawn) of key sections and layouts.

☐ Break down tasks into clear, step-by-step procedural instructions.

☐ Write draft instructions.

☐ Create draft illustrations and figures.

☐ Add warnings and caution statements to mitigate use-related risks.

☐ Include troubleshooting tips and solutions for common issues.

Evaluation and refinement

☐ Conduct one or more formative usability studies to evaluate the design and identify opportunities for improvement.

☐ Refine instructions based on formative evaluations.

☐ Align the instructions' design with the product's branding while prioritizing usability.

☐ Verify that all required regulatory information is included and correctly formatted.

Finalization, verification, and validation

☐ Verify that the instructions meet their established requirements.

☐ Validate the final instructions through usability testing to ensure they enable safe and effective product use.

☐ Perform quality checks on the finalized content, illustrations, and layout.

☐ Prepare the instructions for production and ensure readiness for distribution.

Design evaluation checklist

Content

☐ Is all required regulatory content included and in its required format?

☐ Are all safety-critical warnings and cautions included?

☐ Are all procedural steps included for all necessary tasks?

☐ Is troubleshooting guidance provided to address likely and serious problems?

☐ Is customer support or manufacturer contact information included?

☐ Are cleaning and maintenance instructions included, if applicable?

Organization

☐ Are the instructions organized in a manner that will meet users' expectations?

☐ Is important safety information presented early in the instructions?

☐ Are headings used to convey and clarify the instructions' organization?

☐ Are procedural steps presented in a logical sequence?

☐ Is related content effectively grouped?

☐ Are features, such as a table of contents and an index, included to help users locate specific information?

☐ Are page numbers included, if applicable?

☐ Are cross-references clear and accurate?

Layout

☐ Is content arranged using a consistent grid system?

☐ If the document has multiple sides or pages, is there a clear indication to continue reading?

☐ Are margins and spacing sufficient to prevent crowding of text or images?

☐ Does the layout account for fold lines where applicable?

Language

☐ Is the content at an appropriate reading level for the intended audience?

☐ Are technical terms defined on first use?

☐ Is plain language used throughout?

☐ Are sentences as concise and direct as possible?

☐ Are sentences written in the active voice?

☐ Are the instructions free of ambiguous wording?

☐ Is terminology used consistently?

☐ Are abbreviations, acronyms, and initialisms minimized and/or explained?

Visual design

☐ Is there sufficient contrast between text and background?

☐ Is text sufficiently large to ensure legibility at the expected reading distance?

☐ Do the instructions use an appropriately legible typeface?

☐ Are font styles and sizes used consistently?

☐ Is color used effectively and purposefully?

☐ Do the instructions exclude distracting decorative elements?

☐ Does the instructions' visual design harmonize with the product's appearance?

☐ Do the instructions comply with the manufacturer's brand guidelines, to the extent possible and appropriate?

☐ Would the instructions be sufficiently effective if printed in gray-scale?

Graphics and illustrations

☐ Is the product depicted consistently throughout the instructions?

☐ Are graphics sufficiently large to ensure legibility?

☐ Are important details emphasized appropriately in illustrations?

☐ Are illustrations free of unnecessary or distracting details?

☐ Are the illustration styles consistent?

☐ Are graphics located near their related texts?

☐ Do illustrations clearly distinguish between starting and ending conditions?

Accessibility and inclusivity

☐ Are alternate formats (e.g., Braille, audio, large print) available if needed?

☐ Are graphics and illustrations understandable without relying on color?

☐ Is the language inclusive and respectful?

☐ Are illustrations gender- and ethnicity-neutral or balanced?

☐ Is the document legible and usable in the expected lighting conditions?

☐ Will users with mild or moderate visual impairments be able to read the instructions?

☐ Is adequate space included for the amount of text expansion expected in translated versions?

☐ Do the instructions avoid idioms and culture-specific references?

☐ Are graphics and illustrations appropriate for all intended audiences?

☐ Are measurement units appropriate for all intended markets?

☐ Are text and numbers formatted appropriately for all intended languages?

Warnings and safety information

☐ Are warnings visually distinct from other content?

☐ Do warnings use appropriate signal words (e.g., DANGER, WARNING, CAUTION)?

☐ Do warnings clearly state hazards and consequences?

☐ Are warnings placed where they will likely be seen at the right time?

☐ Are warning symbols and pictorials used appropriately and consistently?

☐ Is important safety information emphasized appropriately?

Production

☐ Is the instructions' paper thick enough to prevent show-through?

☐ Do the instructions fit into their intended package and storage location?

☐ Are the instructions noticeable in the product's packaging?

☐ Are the instructions durable enough for their intended uses and use environments?

☐ Can the document's colors be produced accurately using the intended printing process?

☐ Are digital instructions optimized for use across multiple devices (e.g., tablets, phones)?

☐ Do digital instructions include a search function for faster navigation?

Additional resources

Books

Cutts, M. 2020. *Oxford Guide to Plain English*. 5th ed. Oxford, UK: Oxford University Press.

Evans, P., Sherin, A., and Lee, K. 2013. *The Graphic Design Reference & Specification Book: Everything Graphic Designers Need to Know Every Day*. Gloucester, MA: Rockport Publishers.

Israelski, E., and Muto, W. 2022. *Risk Management: How to Assess and Control Risk Related to Human Error in Products and Systems*. Washington, DC: Human Factors and Ergonomics Society.

Lupton, E. 2010. *Thinking with Type: A Critical Guide for Designers, Writers, Editors, & Students*. Princeton, NJ: Princeton Architectural Press.

Mijksenaar, P., and Westendorp, P. 1999. *Open Here: The Art of Instructional Design*. New York, NY: Stewart Tabori & Chang.

Miller, K. T. 2022. *Graphic Design Fundamentals: An Introduction & Workbook for Beginners*. Durham, NC: K.Y. Design.

Müller-Brockmann, J. 1996. *Grid Systems in Graphic Design: A Visual Communication Manual for Graphic Designers, Typographers and Three Dimensional Designers*. Salenstein, Switzerland: Niggli Verlag.

Samara, T. 2005. *Making and Breaking the Grid: A Graphic Design Layout Workshop*. 2nd ed. Gloucester, MA: Rockport Publishers.

Stocks, E.J. 2024. *Universal Principles of Typography*. Gloucester, MA: Rockport Publishers.

Tondreau, B. 2019. *Layout Essentials Revised and Updated: 100 Design Principles for Using Grids*. Gloucester, MA: Rockport Publishers.

Tufte, E. R. 1997. *Visual Explanations: Images and Quantities, Evidence and Narrative*. Cheshire, CT: Graphics Press.

Wiklund, M., Kendler, J., and Strochlic, A. 2016. *Usability Testing of Medical Devices*. 2nd ed. Boca Raton, FL: CRC Press.

Wiklund, M., Davis, E., and Trombley, A. 2022. *User Requirements for Medical Devices*. Boca Raton, FL: CRC Press.

Zinsser, W. 2006. *On Writing Well: The Classic Guide to Writing Nonfiction. 30th anniversary edition*. New York, NY: Harper Perennial.

Articles and reports

AAMI TIR49:2013/(R)2020. 2013. "Design of Training and Instructional Materials for Medical Devices Used in Non-Clinical Environments." Arlington, VA: Association for the Advancement of Medical Instrumentation.

AAMI TIR 12:2020. 2020. "Designing, Testing, and Labeling Medical Devices Intended for Processing by Health Care Facilities: A Guide for Device Manufacturers." Arlington, VA: Association for the Advancement of Medical Instrumentation.

Breuninger, J. 2019, December. "Legibility: How to Make Text Convenient to Read." UX Collective. Available at https://uxdesign.cc/legibility-how-to-make-text-convenient-to-read-7f96b84bd8af

Cherne, N., Moses, R., Piperato, S. M., and Cheung, C. 2020, July/August. "Research: How Medical Device Instructions for Use Engage Users." AAMI Array. Available at https://array.aami.org/doi/10.2345/0899-8205-54.4.258

"Writing Step-by-Step Instructions." 2022. Microsoft–Learn. Available at https://learn.microsoft.com/en-us/style-guide/procedures-instructions/writing-step-by-step-instructions

Regulatory guidelines and publications

Backinger, C. L., and Kingsley, P.A. 1993 "Write It Right. Recommendations for Developing User Instruction Manuals for Medical Devices Used in Home Health Care." HHS publication FDA 93-4258. Available at https://www.qualysinnova.com/download/files/write-it-right.pdf

"Applying Human Factors and Usability Engineering to Medical Devices — Guidance for Industry and Food and Drug Administration Staff." U.S. Department of Health and Human Services, Food and Drug Administration, Center for Devices and Radiological Health. February 2016. Available at https://www.fda.gov/media/80481/download

"Content of Human Factors Information in Medical Device Marketing Submissions — Draft Guidance for Industry and Food and Drug Administration Staff." U.S. Department of Health and Human Services, Food and Drug Administration, Center for Devices and Radiological Health. December 2022. Available at https://www.fda.gov/media/163694/download

"Electronic Instructions for Use of Medical Devices: Guidance on Regulations." United Kingdom Medicines and Healthcare products Regulatory Agency. Available at https://www.gov.uk/government/publications/electronic-instructions-for-use-of-medical-devices

"Guidance on Medical Device Patient Labeling; Final Guidance for Industry and FDA Reviewers." U.S. Department of Health and Human Services, Food and Drug Administration, Center for Devices and Radiological Health. April 19, 2001. Available at https://www.fda.gov/regulatory-information/search-fda-guidance-documents/guidance-medical-device-patient-labeling

"Human Factors Studies and Related Clinical Study Considerations in Combination Product Design and Development — Draft Guidance for Industry and FDA Staff." Office of the Commissioner, Office of Clinical Policy and Programs, Office of Combination Products, Center for Drug Evaluation and Research, Center for Devices and Radiological Health, and Center for Biologics Evaluation and Research. February 2016. Available at https://www.fda.gov/files/about%20fda/published/Human-Factors-Studies-and-Related-Clinical-Study-Considerations-in-Combination-Product-Design-and-Development.pdf

"Medical Devices; Validated Instructions for Use and Validation Data Requirements for Certain Reusable Medical Devices in Premarket Notifications." A Notice by the Food and Drug Administration. June 9, 2017. Available at https://www.federalregister.gov/documents/2017/06/09/2017-12007/medical-devices-validated-instructions-for-use-and-validation-data-requirements-for-certain-reusable

Walker, M. "Instructions for Use (IFU) Content and Format Draft Guidance for Industry," CDER Prescription Drug Labeling Conference December 4th and 5th 2019. Available at https://www.fda.gov/media/134018/download

Standards

ANSI/AAMI HE75:2009 (R2018). 2018. Human Factors Engineering - Design of Medical Devices. Section 11: User Documentation. Arlington, VA: Association for the Advancement of Medical Instrumentation.

European Commission. 2017. Medical Device Regulation (MDR): Annex I - General Safety and Performance Requirements, Chapter III (Part 1), Requirements Regarding the Information Supplied with the Device, 23. Label and Instructions for Use. Official Journal of the European Union. Available at https://eur-lex.europa.eu/legal-content/EN/TXT/?uri=CELEX%3A32017R0745

IEC 62366-1:2015. "Medical Devices - Part 1: Application of Usability Engineering to Medical Devices." Geneva, Switzerland: International Organization for Standardization.

ISO 13485:2016. "Medical Devices – Quality Management Systems – Requirements for Regulatory Purposes." Geneva, Switzerland: International Organization for Standardization.

ISO 14971:2019. "Medical Devices – Application of Risk Management to Medical Devices." Geneva, Switzerland: International Organization for Standardization.

ISO 20417:2021. "Medical Devices — Information to Be Supplied by the Manufacturer." Geneva, Switzerland: International Organization for Standardization.

Videos

"10 Tips For Writing Effective IFUs For Medical Devices." Informa Content & Media. Available at https://www.youtube.com/watch?v=S2yHi6OS9Fc

"The Anatomy of an IFU." Beyond Clean | Steril Processing Education. Available at https://www.youtube.com/watch?v=Zf9pelBtgG4

"Introduction to Medical Device Labeling Symbols." Emergo by UL channel. Available at https://www.youtube.com/watch?v=1tJo9TO-yuQ

"Risk Basics for Medical Devices." FDA. Available at https://www.youtube.com/watch?v=UCTYCSZ8Tuw

"Forthcoming Revisions to AAMI's IFU Guidance ." AAMI Connect. Available at https://www.youtube.com/watch?v=ecNaLimxlqM

"Writing Instructions and Procedures." Business Communication Essentials. Available at https://www.youtube.com/watch?v=NJ0BiDv9Wa8

Websites

"How to Build the Best User Manual." TechSmith Newsletter. Available at https://www.techsmith.com/blog/user-documentation/

"Design for Reading." Plainlanguage.gov. General Services Administration. Available at https://www.plainlanguage.gov/guidelines/design/

"Designing IFUs and User Manuals for Medical Products - Online course." Emergo by UL. Available at https://www.emergobyul.com/resources/designing-ifus-and-user-manuals-medical-products

US Food and Drug Administration (n.d.) Available at https://www.fda.gov/medical-devices/device-advice-comprehensive-regulatory-assistance/overview-device-regulation

Index

For Product Safety Concerns and Information please contact our EU
representative GPSR@taylorandfrancis.com
Taylor & Francis Verlag GmbH, Kaufingerstraße 24, 80331 München, Germany